# The New Glass Fibre Book

R. H. Warring

**MODEL & ALLIED PUBLICATIONS**
ARGUS BOOKS LIMITED
Station Road, Kings Langley
Hertfordshire, England

Model and Allied Publications
Argus Books Ltd.,
Station Road, Kings Langley,
Hertfordshire, England


First Published 1971
Second Impression 1972
Third Impression 1974
Fourth Impression 1975

ISBN 0 85344 0484

Reproduced and printed by photolithography and bound in
Great Britain at The Pitman Press, Bath

# CONTENTS

# INTRODUCTION

WHEN the first book of this title was planned and produced some ten years ago, glass fibre was a relatively new material for use by amateurs. Whilst the technique used for wet lay-up remains unchanged, and most of the materials are still as originally described, it was felt that a complete re-write, rather than just revision was needed—and so this is virtually a new and up-to-date book on the subject, with the emphasis on the use of glass fibre mat as the normal choice for reinforcement material. At the same time the treatment of the subject has been changed.

A problem with writing any practical book on glass fibre is how to present the subject in the most generally useful manner to readers with different problems, and different ideas on what to make. Rather than describing the making of a number of individual projects, the emphasis has been given to methods, basic techniques and basic forms of mouldings, which can be adapted to individual design requirements. This, it is to be hoped, will give it the widest appeal and *use as a* reference book.

The *basic requirements* and *techniques* are described in Chapters 5, 6 and 7. The reader can refer directly to these for information, or to refresh his memory on "methods". If he wants information on materials, he can go directly to Chapters 3 or 4. For guidance on *design*, he can go direct to Chapter 19. Similarly, for information on other specific subjects he can go directly to the Chapter or Chapters concerned. The book does not have to be read as a whole, although in many cases the individual subjects overlap, or are closely related.

That, in fact, is how the book is intended to be used—as a ready reference to individual subjects on which the reader needs information at any one time to help him with practical work in GRP construction. And if "GRP" seems a little obscure, compared with "glassfibre" or "fibreglass"—a quick read through Chapter 1 will put "GRP" in the picture!

# 1. "GRP"

SYNTHETIC resins are man-made materials, or *polymers* as they are correctly called. They have widespread application as resins (e.g. in paints and adhesives); and also for moulding to produce various solid forms, when the material is generally referred to as a plastic.

All plastics have one thing in common. They lack the strength of metals and other "strong" materials. Certain plastics, like nylon, are of course strong, but they soften and rapidly lose their strength with increasing temperature. This is a characteristic of all *thermoplastic* materials, i.e. polymers which can be changed from a solid to a liquid state by heat, and revert to solid again on cooling. Other types of plastics can be "cured" to set hard by an irreversible process of change. They are only plastic during their formulative stage and once cured remain unaffected by heat, unless this is sufficiently high for them to char or burn. They are therefore far more stable as regards their mechanical properties, but invariably suffer from being brittle materials, unless they are softened by the addition of a plasticiser.

About the only way a resin can be made both strong and rigid is by introducing a reinforcing material in the resin. High strength plastic sheets can be made by introducing layers of paper or fabric in the resin, the result being a plastic laminate. Such sheets cannot be re-formed to other shapes They are "set" in the form the laminate was originally produced—usually flat sheets. They can, however, be re-shaped by sawing, machining, etc., the extent to which this can be carried out depending both on the thickness of the laminate and the resin/reinforcement combination involved. Thus thin decorative laminates are unsuitable for re-shaping, other than cutting and trimming to outline shapes required. Thicker laminate material, like "Tufnol", can be treated as a stock engineering material, like metal.

One material which offers exceptional possibilities for reinforcement is glass fibre. Glass fibre is immensely strong as regards tensile strength, yet lacks rigidity and form on its own. Used in conjunction with virtually any thermosetting resin, the glass could provide the strength that the resin inherently lacks, and the resin the rigidity and form stiffness that glass fibre lacks. This was known for a long time, but the resins available were not really suitable for such a combination. Laminates require heat treatment and pressure to cure properly, which makes sheet materials like paper and fabric much easier to use as reinforcement; and the resins available did not wet glass satisfactorily, so that the bond between glass reinforcement and resin could be far from satisfactory.

With the appearance of cold-setting resins these difficulties disappeared. Not only was the first limitation removed completely, but certain cold-setting resins, and in particular the *polyesters*, readily wetted the resin to form an intimate, perfect bond between the two materials. The real beginning of glass fibre reinforced plastics as practical materials, started with the introduction of polyester resins on a commercial basis in the mid-1940s.

The correct name for such materials is important. They are glass fibre reinforced plastics, usually abbreviated to GFRP, or more conveniently just GRP. They are quite different from ordinary laminates, which are just reinforced plastics or RP's. And the resulting products are not "fibreglass" mouldings! "Fibreglass" is a trade name in this country for glass fibre—and *not* a name for GRP, however much it may be popularly misused in this respect. On the other side of the Atlantic, though, the description "fibreglass" seems to be employed almost universally instead of GRP.

GRP is quite a unique material. It combines the high strength of glass fibre with the rigidity of a thermoset plastic, although the glass/resin combination will not produce a material with the full strength of the glass since the relatively weaker strength of the resin bond will lower the ultimate performance under load. The strength is, in fact, more or less directly proportional to the glass: resin ratio. The fact that the resin can be rendered in liquid form and can be "cured" cold, merely by the addition of a chemical hardening agent, greatly simplifies the making of anything in GRP. Mouldings can be produced merely by draping the glass reinforcement over a suitable mould and the resin virtually painted on, followed by "dabbing" with a brush, or "squeegeeing" with a roller to achieve complete impregnation. The wet lay-up, as it is called, is then merely left to set hard and rigid. The strength properties, and the ease of manipulation are the two outstanding properties of GRP constructions.

A further note on terminology is probably apt at this point. Although glass fibre constructions usually consist of layers of glass fibre reinforcement impregnated with resin, they are not necessarily true laminates. Layers of glass *cloth* used as reinforcement will produce a laminate. Layers of glass *mat* will tend to produce a more homogeneous through-section of random fibres—not layers of reinforcement. In general, therefore, it is better to speak of GRP constructions as *mouldings*, rather than laminates, particularly as "laminate" is the standard description for the aforementioned sheet plastic materials reinforced with paper or fabric. However, GRP mouldings are commonly referred to as laminates.

GRP mouldings are often quoted as being comparable in strength with steel. Such a statement in itself is meaningless, for strength depends on whether a material is being stretched, bent or compressed—the "strength" figure being different in each case. It also depends on the cross-section of the material involved. A wood beam can be as strong as a steel beam, if it is made sufficiently larger in section to compensate for the difference in "beam" strengths of the two materials.

There is also another important factor as far as GRP is concerned. Its strength will depend both on the glass: resin ratio, as previously noted; and the type of glass fibre reinforcement. With some types of reinforcement, such as rovings, strength will be very high in one direction, but less so in other directions. With glass cloth reinforcement this orientation of strength will depend very largely on the weave. With glass fibre mat reinforcement the strength will be more uniformly distributed, but will be lower. Each type of reinforcement will also tend to have different glass: resin ratios, which in turn may be further modified by the manner in which the lay-up is actually done. The strength of GRP, therefore, can only be given in typical figures for different types of reinforcement—not specific figures as can be given in the case of metals, and to a lesser extent woods.

Some comparative strength figures are given in Table 1. Here it will be seen that polyester/glass roving mouldings are, in fact, considerably stronger than steel in tension, but cloth or mat mouldings are not. A glass mat moulding would need to be four times the thickness of steel for equivalent tensile strength. However,

*Table 1*

GRP compared with other Structural Materials

| Material | Glass content % | Specific gravity | Tensile strength lb/sq.in | Modulus of elasticity | Specific strength or strength/ weight ratio |
|---|---|---|---|---|---|
| Polyester/Glass rovings | 70 | 1·9 | 120,000 | $4 \times 10^6$ | 63 |
| Polyester/Glass cloth | 55 | 1·7 | 45,000 | $2 \times 10^6$ | 26 |
| Polyester/Glass mat | 30 | 1·4 | 15,000 | $1 \times 10^6$ | 11 |
| Duralumin | — | 2·8 | 69,000 | $10 \times 10^6$ | 23 |
| Mild Steel | — | 7·8 | 60,000 | $30 \times 10^6$ | 8 |
| Douglas Fir | — | 0·5 | 10,000 | $1·4 \times 10^6$ | 20 |
| Hickory | — | 0·8 | 20,000 | $2·3 \times 10^6$ | 25 |
| Birch Plywood | — | — | — | — | — |
| Mahogany Plywood | — | — | — | — | — |

although the strength figures are not quite as high as many people have been led to expect, where GRP does score particularly is in strength: weight ratio. The same *strength* of glass mat reinforced plastic, for example (four times the thickness of steel), would still *weigh* less than the steel.

In fact, to be strong enough for a comparative job, the GRP moulding would probably not have to be anything like as thick *because* of the more favourable strength: weight ratio. In that case it would be even lighter. The main problem then would probably be to make it rigid enough, so that it did not flex or buckle unduly. This would be done by using stiffeners, or perhaps even sandwich construction. See the chapter on "Design".

The main point is that GRP constructions should not normally be considered on an "equal strength" basis, but designed to utilise the specific properties of the material in the best possible way, i.e. take advantage of its favourable strength:weight ratio whenever possible. In other words, the problem is not how to make a GRP moulding as strong as the same object made in another material, but how to make that object in GRP to perform the same duty. This can result in a substantial saving in material and labour—and total cost. In a majority of cases, though, strength and weight requirements are seldom critical, in which case only commonsense is needed to arrive at a satisfactory specification for a GRP construction, guided by previous experience or general recommendations.

A whole book could, in fact, be written just around the subject of designing in GRP, and the various strength figures attainable. It is rather outside the scope of this present work which is concerned more with practical design and the utilisation of GRP as a material. A lot of useful design and performance data are, however, summarised in the Design chapter and the Appendix, and can readily be referred to when necessary. General lay-up requirements will be found covered in the individual descriptions appearing in the main chapters.

# 2. GLASS FIBRE MATERIALS

GLASS fibre is just what it says—fibres or filaments of glass. Glass itself is a strong material, but brittle. Drawn into filament form it becomes immensely strong and much of its brittle character is lost because it is more flexible in fibre form. The ultimate strength of a glass filament can be as high as 500,000 lb per sq.in., or ten times that of steel. This figure is largely academic, however, for the full strength of glass fibres cannot be realised in practical applications. Nevertheless it serves to show that glass reinforced plastics should—and in fact, do—rate as structural materials.

Glass fibre has been produced commercially for about fifty years. All of the early production was a coarse staple fibre used for thermal insulation. This material is generally unsuitable as a reinforcement for plastics. Later techniques produced continuous specifically sized filaments, bundled together immediately after drawing to form strands. These strands can be twisted together, like loosely laid rope, to produce *rovings*, or into *yarn*. Further products can be produced from either of these forms (see *Fig. 2.1*)

Yarns are woven to produce *glass fibre cloths*. These are woven in continuous lengths, usually 36 in. wide, in a variety of different weaves. Some of the common weaves are shown in *Fig. 2.2*. Glass cloth is durable and strong, but both its strength and smoothness depends to a very large extent on the weave. It is also a relatively expensive material to produce.

Continuous filament rovings usually consist of 60 to 120 strands and may also be woven into cloth (*woven rovings*). Alternatively rovings are chopped into lengths of about 2 in. and arranged in random form, in layers, to make glass fibre mats, generally known as *chopped strand mat*. The layers of mat are bonded together with a chemical binder, chosen to dissolve in the resin used in GRP lay-up.

Like cloths, glass fibre mats are manufactured in continuous lengths, with the thickness designated by the weight per sq.ft. Usual weights are 1 ounce mat, $1\frac{1}{2}$ ounce mat and 2 ounce mat. As a very rough guide the *finished* thickness of resin-impregnated mat used in GRP laminates is about $\frac{1}{32}$ in. per ounce weight. In other words, a single layer of $1\frac{1}{2}$ ounce mat would produce a finished thickness of about $\frac{3}{64}$ in.

It should be noted that older type mats were produced with a binder which was compatible with, rather than dissolving in, the resin. This entailed the use of considerably more resin to produce satisfactory wetting out, and generally less satisfactory laminates. The material also bulked more, yielding about twice the thickness given above, due almost entirely to the higher resin content, or lower glass:resin ratio. Strength was also lower because of the lower glass:resin ratio.

Various other forms of mat may be encountered, namely:

*Needleloom*, which is composed of 2 in. chopped strands held together mechanically by perpendicular strands "needled" through the backing material. It is stronger than ordinary chopped strand mat in its dry form and drapes rather more readily. Its main application is for lay-ups where some pressure is applied.

*Coiled mat* is produced from continuous strands, laid random fashion with a suitable binder, and produces a laminate intermediate in strength

FIG. 2·1

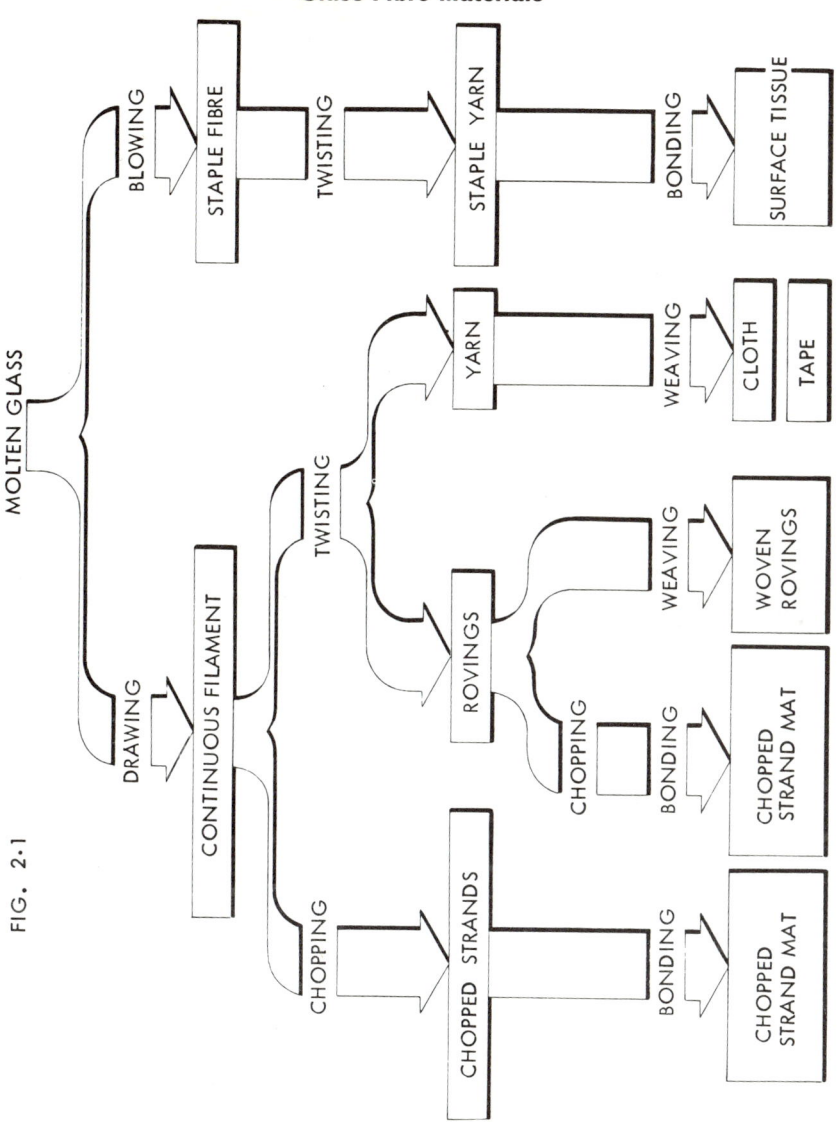

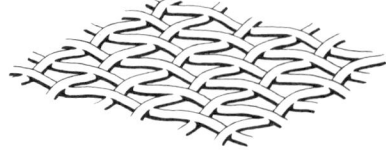

Drawing, to much enlarged scale, of plain weave, strands alternately crossing over and under.

Drawing of twill weave in woven cloth for general purpose use.

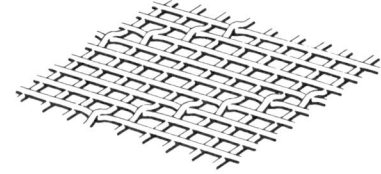

Drawing of satin weave. Satin weave produces a smooth cloth.

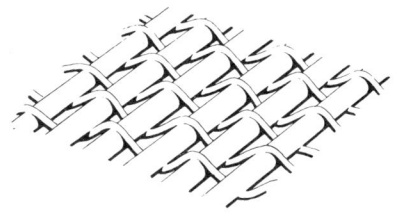

Undirectional weave where the cloth is appreciably stronger in one 'preferred' direction.

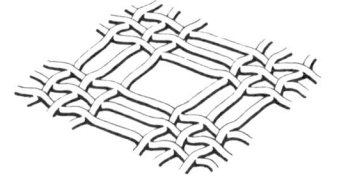

Mock Leno weave, showing the individual strands grouped in threes and interwoven.

FIG. 2.2

between chopped strand mat and woven cloth.

*Surfacing mat* is a very thin form of mat, usually about 0.012 in. thick, is normally called glass tissue, or *glass fibre tissue*. It has very little strength and is used mainly as a gel coat "support".

*Diamond mat* is an older form, more correctly classified as a "cloth" or "semi-cloth". It is produced from continuous strands arranged in a diamond pattern with an angle of 15 degrees between intersecting strands, and bonded together with a suitable chemical binder. It is stronger and easier to handle than chopped strand mat in its dry state, and also produces a laminate with high uni-directional strength. It is little used these days, however, and may not even be obtainable.

Glass is also produced in the form of *staple fibres*, by blowing instead of drawing from the molten glass. Stable fibres can be twisted together to form relatively "soft" yarns with good bulk rather than high strength. They also drape well and have the affinity to absorb a high proportion of resin and wet out readily. Staple yarns themselves are not widely used for reinforcement. However they can be formed and bonded to produce thin sheets or tissues of low strength, known variously as overlay mats, surface tissues and glass tissues.

## Glass Fibre Materials

Choice of reinforcement material for GRP laminates depends on the particular requirements of the job, cost, ease of working, and other factors. The following is a general guide.

1. *Glass fibre mat*. This is the usual material employed for GRP laminates since it is easy to shape (necessarily easiest to handle), costs considerably less than glass cloth, and

with modern soluble binders is readily wetted out. Glass:resin ratios of a similar order to those achieved with cloth are readily possible.

2. *Woven glass cloth*. These generally produce the strongest laminates, but are not always easy to wet out thoroughly. The greatest tensile strength is obtained by using the thinnest cloths with the closest weave, which can aggravate this particular problem. As a consequence, interlaminar adhesion can be poor locally. Inter-layer bonding is usually improved by using thicker cloths and fewer layers.

3. *Woven roving*. These cloths are obtainable in various thicknesses—e.g. in weights from about 1 ounce per sq. ft upwards. They do not have the same strength as woven cloths but drape well and wet out fairly easily. Since they cost less than woven cloths they are often used as a reinforcement layer or layers in conjunction with glass fibre mat, or as a simple means of adding bulk to increase stiffness.

4. *Rovings*. These again are normally used for reinforcing glass mat layers, and also to add bulk and improve stiffness. They are widely used for strength purposes in boat hull construction.

5. *Glass fibre tissue*. This was originally developed to hide the pattern of glass cloths when laid up in matched metal moulds, and also to be used as a "cushioning" layer between the gel coat and the main reinforcement layers in any moulding or lay-up produced in a female mould. For general use it has two main applications—

(a) As a finishing layer on a chopped strand may lay-up to cover up the coarse glass pattern of the mat before the gel coat.

(b) As a finish applied over a set

gel coat. Again this will prevent any glass pattern appearing in the finished surface, and it also provides some reinforcement for the gel coat.

6. *Glass fibre tape*. This is simply woven cloth in tape form, and is the most convenient material to use where narrow strips of reinforcement are required for spiral bindings. It is a strong material, readily handled, and easy to use for small repair jobs, etc. Tape is correctly specified by width and thickness.

### Pre-impregnated Glass Fibre

Glass fibre cloth and mat is also produced uniformly impregnated with a resin-catalyst mix which can be activated by heat. In other words, the material is complete in itself for lay-up work which can be cured by heat and pressure. In its original state the material is slightly tacky, which can help in laying up. It is, however, only used for hot press mouldings and generally unsuitable for amateur work because both heat and pressure are required to cure. Pre-impregnated glass fibre is known variously as "pre-preg", sheet moulding material, etc.

### Types of Glass

Ordinary glass, or soda-glass, can vary considerably in composition—mainly as regards its alkali content, which affects its resistance to moisture and chemical attack. A low alkali content implies high resistance to moisture, whilst a high alkali content implies good resistance to attack by acids. Soda-glass may have an alkali content of anything from 8 to 15%, so it does not have particularly good resistance to water, but is strongly resistant to attack by most acidic solutions and chemicals. It is thus well suited for general applications and is specifically known as "A" grade glass.

Where extreme resistance to moisture is required, a boro-silicate glass is to be preferred. This has an alkali content of less than 0·8% and for all practical purposes can be regarded as completely unaffected by water. It is more expensive to produce than soda-glass, and is generally known as "E" or "Electrical" grade glass. Glass fibre mat (or cloth) made from "E" grade or boro-silicate glass would, therefore, normally be chosen for mouldings that are continually exposed to water, such as boat hulls, garden pool liners, etc., although this is not necessarily always followed in practice. "E" grade glass cloth or mat, in fact, may be more difficult to obtain. On the other hand, "E" grade glass would not be chosen for a job requiring maximum resistance to chemical attack.

Normal supplies of glass cloths and mats are made from soda-glass fibres or "A" glass. These are suitable for all general applications as the alkali content is usually reasonably controlled, offering a good compromise between good water resistance and good resistance to acids and chemicals. Glass, as is well known, is a pretty inert substance, and in a GRP laminate it will normally be the resin which finally determines the resistance of the finished laminate to chemical attack.

# 3. RESINS

QUITE a number of different resins could be used for GRP constructions. Practically all the work done with this material, however, employs an *unsaturated* polyester resin. "Unsaturated" really means that the resin is capable of being "cured" from a liquid to a solid state, this being brought about by dissolving the polyester in a suitable monomer, usually styrene. A poleyster resin of this type can be made to set in the form of a hard and permanent solid by the addition of a catalyst, and satisfactorily moulded without the use of pressure. In this state the resin is said to be cured, or polymerised.

Curing with just the addition of a catalyst also requires heat to complete the process. However the resin can be made to set at room temperatures or "cold cure" by the further addition of an accelerator. This is obviously preferable, for it considerably simplifies the process. The two types of cure are essentially the same. With cold curing, heat is actually produced exothermically (or internally, as far as the resin is concerned), and so the end result is exactly the same. Resin plus catalyst, hot-cured, has the same characteristics as resin plus catalyst plus accelerator, cold-cured.

A polyester resin is not a single material. There are many different types of these resins, each with different characteristics. These may range from strong, hard solids when cured, to quite flexible materials. The majority are formulated for GRP lay-up work. Others are specially formulated for casting, surface coatings, stopper compounds, nut locking or thread sealants, and even as a mortar for concrete constructions. Some idea of the variety of different types and applications is given in Table 2, and this particular range does not include

types specially developed for decorative castings and encapsulation. But all are unsaturated resins, capable of being cured. *Saturated* polyesters are quite a different material, and cannot be cured in the same way—e.g. Terylene is a saturated polyester.

All suppliers normally offer a range of polyester resins for GRP work. These may or may not be pre-mixed with accelerator. Pre-accelerated resins are normally the most convenient to use since the accelerator proportion is already adjusted to give the most suitable gelling and hardening characteristics, once the catalyst is added to start the curing reaction at room temperature. If a "straight" resin is used, then accelerator and catalyst must be added separately, the order of mixing depending on the particular application involved.

A "straight" resin will have a natural tendency to polymerise or harden naturally, although this process will be very slow. Stored in the dark, or in a close container, and at a reasonably low temperature, the resin will remain sufficiently liquid to be usable for up to twelve months. Exposed to light (and especially sunlight), or heat, the shelf life of the resin will be drastically reduced. It could be as little as a week if left in an open or glass container in sunlight. As a simple test of the condition of the resin it can be considered usable as long as it remains liquid enough to be poured from its container. The pre-mixing of accelerator with resin does not affect its shelf life, so there is no difference between a "straight" resin and a pre-accelerated resin as to the time they will keep in stock and remain usable.

With a pre-accelerated resin, addition of the catalyst (or *hardener*, as it is often called) will immediately start the cold-curing process, causing the

*Table*

14

Some Types of

| Type LEGUVAL RESIN | Typical characteristics | | | | | | Applications | | | | | | |
|---|---|---|---|---|---|---|---|---|---|---|---|---|---|
| | | | | | | | | | moulding | | | | |
| | heat deflection strength | chemical resistant | flame resistant | flexible resin | contains amine accelerator | light stabilized | hand lay-up | resin/fibre spraying | unheated mould | heated mould | sheet moulding compound and moulding compound | translucent sheet and dome lights | pipe, containers |
| N 30 | | | | | | | ○ | ○ | | ● | ● | | ○ |
| N 50 | | | | | | | ○ | ● | | | | | ○ |
| N 50 S | | | | | | ● | ○ | ○ | | | | ● | |
| W 16 | ● | | | | | | ● | ● | ○ | | | | ● |
| W 16 T | ● | | | | | | ○ | ● | | | | | |
| W 20 | ● | | | | | | | | | | ● | | |
| W 35 | ● | ● | | | | | ○ | ○ | | ○ | | | ○ |
| W 41 | ● | ● | | | | | | | | ○ | ○ | | ○ |
| K 41 | ● | ● | | ● | | | | | ○ | | | | ○ |
| F 30 | ● | | ● | | | | ○ | ○ | ○ | ○ | | | |
| F 30 S | | | ● | | | ● | | | | | | ○ | |
| F 33 | | | ● | | | | | | | ○ | ● | | |
| F 35 | | | ● | | | | ○ | ○ | ○ | ○ | ● | | |
| E 60 | | | | ● | | | | | | | | | |
| K 70 | | | | ● | ● | | | | | | | | |
| E 81 | | | | ● | | | | | | | | | |
| E 90 | | | | ● | | | | | | | | | |
| K 25 R | | | | | ● | | ○ | | | ● | | | |
| K 26 | | | | | ● | | ○ | | | ● | | | |

| boat hulls, housings enclosures | putties, filling compositions | putty compounds | mortar screeding | polyester/ concrete | polyester artificial stone | Specific properties |
|---|---|---|---|---|---|---|
| ● | | | | | | high mechanical strengths |
| ● | | | ● | ● | | standard type |
| | | ○ | | | ● | light stabilized type for translucent sheet |
| ● | ○ | | ○ | ● | ○ | standard type with improved heat deflection strength |
| | | | | | | thixotropic |
| | | | | | | for sheet moulding compounds and moulding compounds |
| ○ | | | ○ | ○ | ○ | good chemical resistance; good adhesion to UPVC; neopentyl glycol type |
| ○ | | | | | | high chemical resistance; bisphenol A type |
| ○ | | | | | | Leguval W 41 with accelerator content; bisphenol A type |
| ○ | | | | | | flame resistant; HET acid type |
| | | | | | | flame resistant, light stabilized for translucent sheet, HET acid type |
| ○ | | | | | | flame resistant, containing chlorinated paraffin wax, based on Leguval N 30 |
| ○ | | | | | | flame resistant, containing chlorinated paraffin wax, based on Leguval W 20 |
| | | ● | | | | flexible resin of high tensile strength; neopentyl glycol type |
| | ● | ○ | ○ | | | for filling compounds, tackfree surface with cobalt, contains accelerator |
| | ○ | | | | | highly versatile flexible resin |
| | ○ | ● | ● | | | flexible resin with relatively low water absorption |
| ○ | ● | ○ | ○ | ○ | | standard type containing accelerator, for press moulding in unheated moulds, filling compounds, putties and mortars |
| ○ | ● | ○ | ○ | ○ | | standard type containing accelerator, for press moulding in unheated moulds, filling compounds, putties and mortars |

resin to thicken and gel, and finally set hard. The resin has a usable pot life over the period during which it is thickening, but still remains workable. The gel time is governed mainly by the amount of *accelerator* present. The less the accelerator the longer the gel time. Cutting down the amount of catalyst will also increase the gel time, but here there is a danger of having insufficient catalyst present, which can lead to an undercured moulding. With a pre-accelerated resin, therefore, it is less advisable to try to control gel time by adjusting the amount of catalyst used, but work strictly to recommended proportions of catalyst. If faster, or slower, gel times are required, then a pre-accelerated resin giving these particular characteristics can be used.

There is, however, another way of adjusting the gel time. This is by using a different type of catalyst. Thus with a pre-accelerated resin the following alternatives may be offered by various suppliers:

(i) Resins with different proportions of accelerator (or different accelerators) to give (a) normal gel times; (b) faster curing for colder weather use, or emergency repairs; (c) slower curing for increased pot life in hot weather. Catalyst proportions would be specific in each case.

(ii) A single resin which can be used with different catalysts to give different gel and hardening times.

With a "straight" resin, the choice of accelerator and catalyst is under the control of the user, who can thus adjust the mix to suit his particular requirements. However, in this case there is more chance of going wrong. The system is more flexible though, for there is the further choice of—

(i) Mixing the *catalyst* with the resin

first, which will give a pot life of some 20 hours or more under normal conditions, adding the accelerator in the proportion required for making up the mix for immediate use.

(ii) Mixing the accelerator with the resin, and adding the catalyst in the proportion required for immediate use. This is virtually the same as using a pre-accelerated resin, except that the user has control over the type and proportion of accelerator used.

It is important to note that *accelerator and catalyst should never be mixed together directly*. This could result in a violent, even explosive reaction. One or other must *always* be added *to the resin first*.

### Types of Resins

The following are the main types of resins generally available for amateur use, normally sold together with matching accelerator (if not pre-mixed with the resin) and catalyst (hardener).

*General Purpose Resin*—which may also be called "laminating" resin. This is suitable for all kinds of GRP work, and the price is generally moderate. It is most widely supplied in pre-accelerated form, with a standard catalyst. An additional catalyst may also be available to produce more rapid hardening at low temperatures. This is normally called a "winter additive" or similar, and is added to the main catalyst in specific proportions when necessary.

*"Marine" or "High Duty" Resin*—this is (or should be) a superior quality resin with high water resistance, higher strength and particularly good adhesion to wood. It is particularly suitable for mouldings exposed to water, such as boat hulls, hull sheathing, garden pools, etc., and because these applications frequently involve vertical mould

surfaces, a thixotropic agent may also be included in the resin. Price is roughly twice that of a general purpose resin.

*"Non-Sag" or Thixotropic Resins*—these may be of various types, the main feature being the incorporation of an additive which prevents the resin running and thus draining off vertical surfaces in a lay-up. Ordinary resins can also be made "non-running" by the addition of suitable thixotropic agents.

*Gel Coat Resins*—these are specially formulated to give tough, resilient films and are intended only to be used for gel coats. A disadvantage of many resins of this type are that they tend to remain tacky unless air is completely excluded from the surface when curing—e.g. if the gel coat is exposed in the mould, it would have to be covered with cellophane sheeting.

*Rapid Resins*—these are pre-accelerated resins where the accelerator used, and matching catalyst, produce fast gel and hardening times. They are intended mainly for emergency repair work, but can also be used for general work in colder weather, provided the mouldings involved are not too large in size.

*Clear Resins*—these are also known as "rooflite" resins, translucent resins, etc., and are light-stabilised resins intended for use in the production of translucent rooflight sheeting, decorative panels, etc. They are formulated to give optimum optical properties with freedom from discoloration with age (all normal GRP mouldings tend to yellow slightly with age, although this effect is not usually very noticeable).

*Casting Resins*—these are specially formulated for casting work (see Chapter 8). Their properties are "tailored" to the characteristics required, e.g. good optical properties for resins used for decorative work or embedding botanical or zoological specimens; good dielectric properties for resins used to encapsulate electronic components. Low volumetric shrinkage is also important in a casting resin.

*Flexible Resins*—these are used for blending with a more rigid resin to vary the "elastic" properties, and thus control the resilience of the finished GRP moulding.

*Plasticised Resins*—these are softened resins, produced by the addition of a plasticising agent such as dimethyl phthalate. Again these impart flexibility to the resin, but are employed only in special circumstances. These include the formulation of resins as thread locking compounds, and resins for the production of large castings. Plasticised resins should never normally be used for introducing resilience into GRP mouldings as there is always the distinct possibility of the plasticiser leaching out.

*Fire-retardant resins*—these are usually of chlorinated type and may be mixed with "straight" resin to improve its resistance to flame, or make it self-extinguishing.

*Special-purpose Resins*—these are formulated to enhance specific properties and are not normally generally available. They are intended for commercial productions calling for properties not met by standard resins.

*Sealers and Varnishes*—these are simple coating resins, intended to be used on finished mouldings either to produce a better surface appearance, or seal the surface, or both. They may also contain air-inhibitors to avoid surface tackiness. The two main types are polyester resin-varnish and polyurethane resin-varnish. The latter has good adhesion to polyester, with the advantage that it sets to a hard, abrasion resistant coating which can be ground and polished to a mirrorlike surface.

Sealers and Varnishes, of course, also have an application for sealing the surface of moulds made from porous materials, such as Plaster of Paris, wood, cardboard, etc.

*Accelerators*

Quite a number of different chemical compounds are effective as accelerators in promoting cold-setting of polyester resins in the presence of a catalyst. Some have only a limited use (such as tin, vanadium and zirconium salts and certain ammonium compounds). Others are highly reactive, and normally used. These include metallic soaps (usually cobalt soap); and tertiary aminies (such as dimethyl aniline). There is probably not much to choose between the two, except that an amine accelerator tends to produce a yellow discoloration in time. If the two types of accelerator are used together, however, much faster gelling and setting times can be realised. This method is used to produce a "rapid" accelerator.

*Catalysts (Hardeners)*

The catalysts used with polyester resins are almost invariably organic peroxides. These are unstable on their own—and can even explode. They are therefore normally supplied dispersed in a plasticiser in the form of a paste or a liquid, or in dry powder form mixed with an inert filler.

Catalysts should always be handled with care, especially if relatively large quantities are involved. They are all irritating to the skin and can cause burns unless washed off immediately. Injury can be more serious if catalyst is splashed into the eyes. Immediate treatment in such cases is to wash out the eyes continuously with plain water or weak bicarbonate solution.

Rags or waste used to mop up catalyst, or on which catalyst has been spilt should be damped down with water and thrown away in a dustbin. They should not be left lying around as they may self-ignite through spontaneous combustion.

The chief catalysts are:

(i) methyl-ethyl-ketone-peroxide— which is probably the most reactive type, normally produced as a liquid.

(ii) cyclohexanone-peroxide—which is less reactive, but more stable. It is normally prepared in paste form.

(iii) benzoyle-peroxide — which is again very reactive and is normally available in either paste or powder form.

For amateur work a liquid catalyst (hardener) is generally preferred since it is easier to measure accurately, and disperse uniformly through the resin by stirring. Powdered catalysts, mixed with inert fillers, are used for fillers, mixes and body stoppers, in conjunction with a pre-accelerated resin (usually of the "rapid" type).

*Proportions*

The setting action of the resin is initiated by the *catalyst* or hardener. The accelerator merely takes the place of heat—i.e. in the absence of the accelerator the catalysed resin can be cured by heat; or in the presence of accelerator the catalysed resin will cure at normal room temperature. The accelerator must, therefore, be present in sufficient quantity to activate the catalyst, if complete cold curing is to be produced in a short period of time. If there is a small deficiency of accelerator the gel time will be increased, whilst a larger deficiency of accelerator will result in undercuring, or very slow hardening. The proportion of accelerator, therefore, is not critical, provided there is enough. Usually this is between 1 and 4 % of the resin weight,

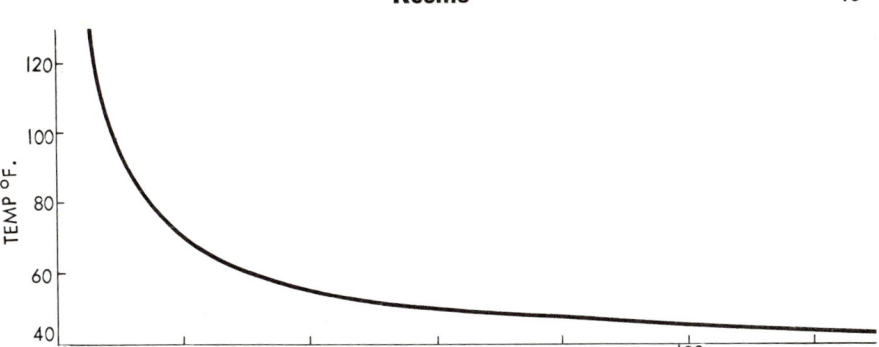

GEL TIME–MINUTES FIG. 3·1

with the normal "strength" of the accelerator as manufactured (e.g. typically a dilution giving 1 % metal in the case of a cobalt soap).

Catalysts are usually formulated on a 25–50 % peroxide "strength", which yields a similar proportion requirement to the accelerator. Recommended proportions usually range between 1 % and 4 %, depending on the type of catalyst, and the gel time required. Very roughly, doubling the amount of any catalyst from 1 to 2 % will halve the gel time, but a further increase in catalyst will produce a more rapid increase in hardening. The gel time will not be greatly affected by increasing the catalyst above about 4 %, however, as shown in *Fig. 3.1*. Lack of catalyst, on the other hand, can lead to an incomplete cure.

*The Cold-Curing Reaction*

From the moment the catalyst is activated (either by heat or contact with the accelerator), the resin starts to set. Setting takes place over four definite stages:

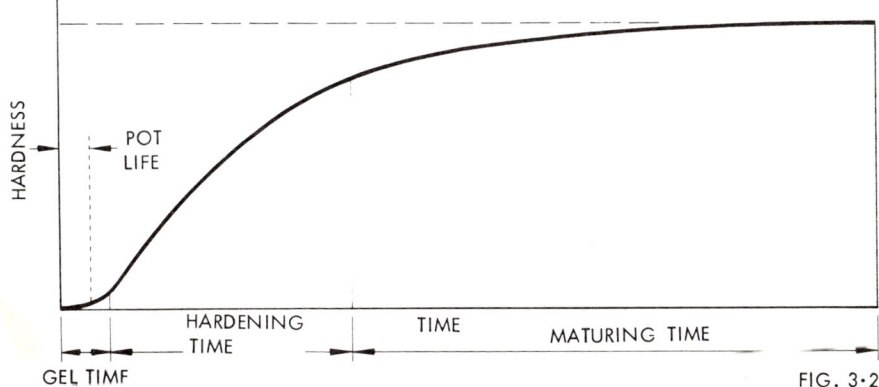

FIG. 3·2

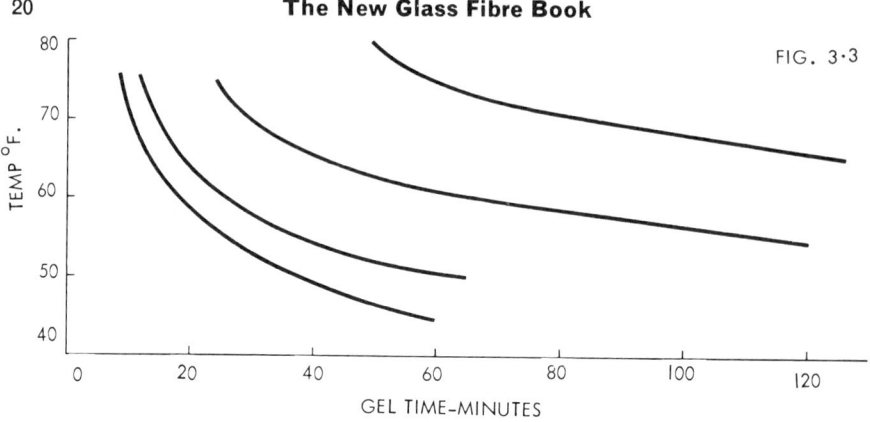

FIG. 3·3

GEL TIME–MINUTES

GEL TIMES FOR A TYPICAL POLYESTER COLD-CURING RESIN

1. *Pot life time*—during which the resin still remains in a workable liquid form although it is continuing to thicken.

2. *Gel time*—or the time taken for the resin to set to a soft gel.

3. *Hardening time*—which is the further time taken for the resin to become hard enough for the moulding to be removed from its mould.

4. *Maturing time*—which is the further period of time over which the moulding will continue to gain hardness and,

eventually, complete stability. When fully matured the moulding will have its maximum strength, hardness, chemical resistance and stability.

These four stages are illustrated in *Fig. 3.2*.

Pot life and gel time are strictly related to the activity of the catalyst. This is governed both by the proportion of the catalyst and the ambient temperature. For a given proportion of catalyst, the higher the air temperature the shorter will be the pot life and ge time of the resin, as shown in *Fig. 3.2*

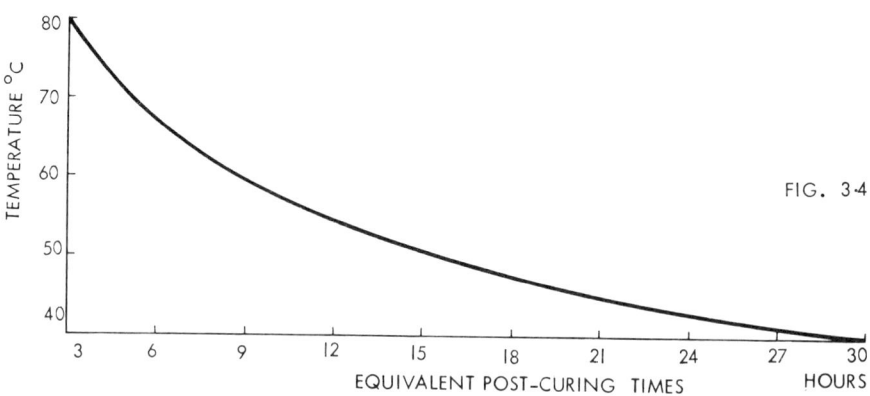

FIG. 3·4

EQUIVALENT POST-CURING TIMES      HOURS

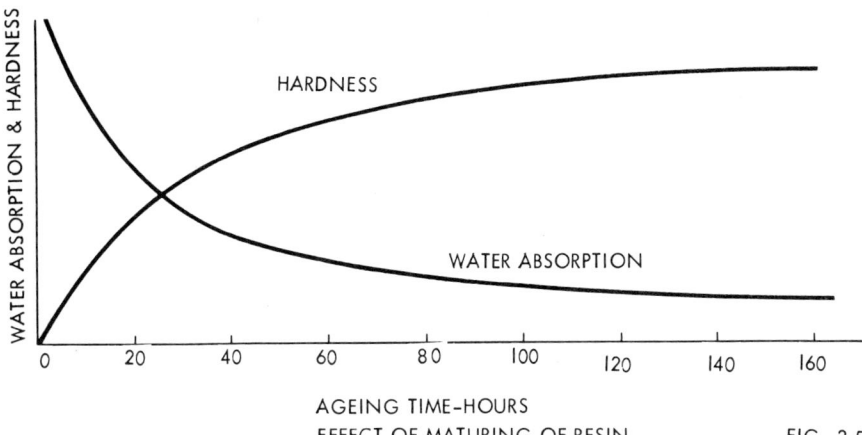

AGEING TIME-HOURS
EFFECT OF MATURING OF RESIN                    FIG. 3.5

This diagram also shows the general effect of varying the proportion of catalyst to control the gel time at different ambient temperatures. These curves can only be used as a rough guide, however. They are specific to a particular resin-accelerator-catalyst combination. Most proprietary resin-hardener combinations do, however, give data on pot life and gel times for different temperatures, and details of how these can be adjusted, if necessary, or possible.

The *hardening time* can vary a lot depending on the size and thickness of the moulding, and also the proportion of resin present. It will also again be affected by the air temperature. The warmer the air the more quickly the moulding will harden off. With a small moulding, and reasonably warm air, hardening time may only be an hour, or even less. With a large moulding it may need 12 or even 24 hours before the moulding is hard and rigid enough to be removed from its mould without

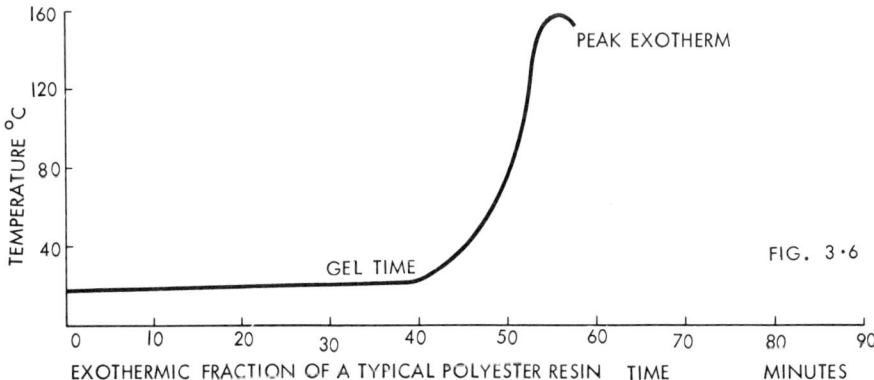

EXOTHERMIC FRACTION OF A TYPICAL POLYESTER RESIN    TIME    MINUTES

fear of it sagging slightly, or becoming distorted in the process.

The majority of general purpose GRP mouldings can be considered suitable for use 24 hours after their hardening time has elapsed. Where the application is more critical, such as GRP boat hulls, water or petrol tanks, etc., or articles which must have good chemical resistance, a maturing time of several days, or even weeks may be necessary. It is recommended that a GRP fuel tank, for example, should be allowed a maturing time of at least one month, to be sure that the material will be fully stable when put into use.

For rather less critical applications, maturing time can be reduced to a matter of a few days by storing the moulding in a warm atmosphere. *Fig. 3.4* shows the effect of atmospheric temperature in terms of equivalent maturing times. Thus at a temperature of 80°C, for example, the maturing time is only one tenth that which would be necessary at 40°C. It is important not to try to accelerate maturing time with too much heat, however, and if a moulding is to be stored in a temperature much above 40°C it should first be allowed to mature naturally, at ordinary temperature, for 24 hours. If not, it may show signs of warping or distortion.

Some idea of the effect of maturing time on the properties of the GRP moulding can be obtained from *Fig. 3.5* which shows how resistance to water and actual hardness improve with age.

*The Exothermic Effect*

It has already been mentioned that the cold-setting of the resin does, in fact, produce heat. This heating, which causes a rise in temperature in the GRP, does not take place until the resin has passed the gel state and is almost set—see *Fig. 3.6*. This effect is not normally noticeable, or significant, with the usual run of GRP work, but it can be important where a large bulk of resin is involved, such as in thick mouldings. To avoid excessive heating which could expand and distort the mould itself, it may be necessary to apply water cooling to the mould in such cases; or stop the lay-up after several layers of reinforcement have been added and let these harden before proceeding with the lay-up.

What will be noticed with ordinary mouldings, is that the thicker the section the more rapid the hardening period is likely to be, because of the internal heat developed. Castings may need special care in order to ensure that the actual temperature of the set resin does not reach levels which could be damaging to objects which are being encapsulated.

*Getting the Proportions Right*

The majority of amateur constructions are done with pre-accelerated resins, leaving only the catalyst (hardener) to be added in recommended proportions. These proportions may range from 1 to 4%. These proportions can be read as the *weight* of catalyst (hardener) required to mix with a given *weight* of resin. Where the catalyst is a liquid or paste, this involves translating weight into terms of liquid measure.

Where small quantities of resin are involved, the catalyst (hardener) can be added drop by drop, using the following count:

5   drops   equals   1%   catalyst (hardener) per 1 ounce of resin

10   drops   equals   2%   catalyst (hardener) per 1 ounce of resin

15   drops   equals   3%   catalyst (hardener) per 1 ounce of resin

20   drops   equals   4%   catalyst (hardener) per 1 ounce of resin

This still requires knowing the *weight* of resin to be used, but the proportion of resin required to match a particular glass fibre reinforcement can only be accurately determined by *weight* (see Chapter 6). You must, therefore, also have facilities for weighing.

For example, if the job calls for 2 ounces of resin, this can only be measured out by weighing an empty glass jar or similar container and then pouring resin into this container until the weight indicated has increased by 2 ounces. Alternatively you may be able to buy graduated measures (or one may be supplied with the resin), which shows the amount (volume) of resin to be poured in to make a certain weight.

Having obtained the 2 ounce measure of resin, the mix is to be prepared by adding, say, 3% catalyst (Hardener). Now 3% is 15 drops per 1 ounce of resin; and so 3% for the 2 ounces would be 2 × 15—30 drops.

For quantities of resin more than a few ounces it is more convenient to use a liquid measure to determine the amount of catalyst (hardener) to be added. A measuring beaker is the answer here, normally graduated in *millilitres*. The following measures then apply:

5 millilitres equals 1% catalyst (Hardener) per 1 pound of resin

10 millilitres equals 2% catalyst (Hardener) per 1 pound of resin

15 millilitres equals 3% catalyst (Hardener) per 1 pound of resin

20 millilitres equals 4% catalyst (Hardener) per 1 pound of resin

The same measures, of course, can apply if accelerator is to be added to the resin—or, in fact, any other additive. In the case of powdered accelerators or catalysts, measures should be done by weights, when Table 3 can be used

*Table 3*

Proportions (weight in ounces, except where noted)

| Weight of Resin | % Catalyst or Accelerator | | | |
|---|---|---|---|---|
| | 1 | 2 | 3 | 4 |
| 2 oz | 10 drops | 20 drops | 30 drops | 40 drops |
| 4 oz | 20 drops | 40 drops | 60 drops | 0·16 (⅛) |
| 8 oz | 40 drops | 0·16 (⅛) | ¼ | 0·32 (⅓) |
| 1 lb | 0·16 (⅛) | 0·32 (⅓) | ½ | 0·64 (⅔) |
| 2 lb | 0·32 (⅓) | 0·64 (⅔) | 1 | 1·3 |
| 5 lb | 0·64 (⅔) | 1·3 | 2½ | 3·2 |
| 10 lb | 1·3 | 3·2 | 5 | 6·4 |

# 4. FILLERS, PIGMENTS AND PARTING AGENTS

A FILLER is an inert substance, usually a powder, which may be added to liquid resin to give it more bulk and also "through colour". Because resin is relatively expensive, the use of fillers can reduce the cost of a GRP moulding of required thickness and at the same time somewhat increase its compressive strength. However it is likely that the overall strength will be reduced if the proportion of fillers used is too generous. Also the introduction of fillers in the resin can make it more difficult to judge whether or not a lay-up is suitably consolidated with the exclusion of air bubbles, or introduce local weaknesses if the filler is not completely and uniformly mixed with the resin.

Whilst fillers were once widely favoured as giving a more solid appearance to the GRP moulding, because of the through colouring, the use of fillers in high class functional mouldings is now usually avoided. Such mouldings are characterised by a translucent appearance, with colouring applied only in the gel coat (and perhaps a flock spray coating on the other side). To the uninitiated the moulding may appear to lack "body" or thickness as a consequence, but this is an illusion. But more important still, and defects in the lay-up will show, because of the translucence of the whole moulding.

Fillers can, however, be most useful where "body" is required, or through-colouring is particularly desirable. Special types of fillers may also be used to improve both tensile and impact strength. These would be fibrous type fillers, such as asbestos fibre, chopped glass fibre or macerated fabrics.

A further use of fillers is for rendering the resin in the form of paste rather than a liquid. In this form it can be used as a filler for car body repairs, etc., without the addition of further reinforcement. The usual combination is a pre-accelerated resin and separate powder comprising filler mixed with powdered catalyst. The two are mixed in suitable proportions to provide the necessary stiff paste for filling and stopping work, setting hard by cold-curing in the usual way. Alternatively, resin and filler may be mixed to provide one paste, catalyst being supplied as a separate paste. In this case both products can be tubed for convenience of use.

Simple resin fillers of this type may be based on rigid resins, setting as hard as a normal GRP moulding; or on flexible resins to give a slightly elastic solid when set. The latter have the advantage that they are easier to re-work for flatting down, etc. Other types of fillers may have reinforcement introduced, as well as fillers, and may be suitable for the production of moulded shapes. Casting resins may also incorporate fillers, for reinforcement, colouring or other decorative effects.

A basic requirement of any filler is that it should be inert and have no inhibiting effect on the resin cure. Also the filler material should not be subject to ageing or deterioration under the service conditions involved for the final product. The usual choice is a mineral filler, such as precipitated chalk, China clay, talc or whitening. These materials are satisfactory, provided they are pure and very finely powdered. They will then mix uniformly with the resin, with the effect of increasing its viscosity, i.e. tending to make it become more paste-like. The higher the proportion of filler the more stiff the resin becomes for working. This effect is even more marked with the modern surface-

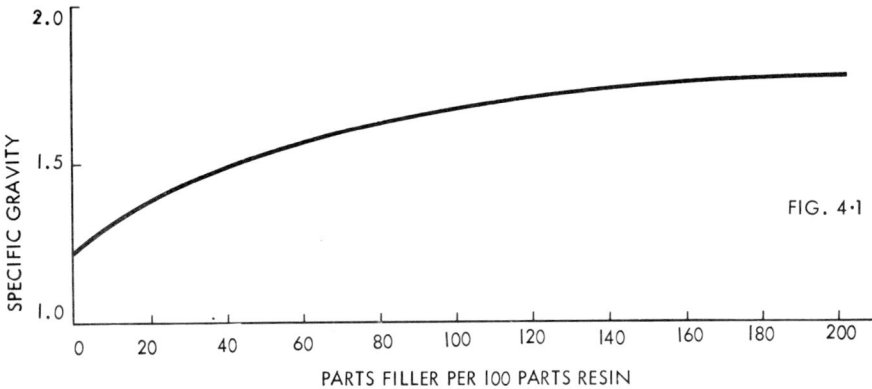

FIG. 4·1

INCREASE IN WEIGHT WITH FILLERS

treated calcium carbonate fillers, particularly crystalline types, which have become the preferred choice. These also improve the impact strength of the moulding, as well as producing a harder, more scratch resistant surface. This type of filler is also used where mouldings in light or bright colours are required, without going to the expense of titanium dioxide or other similar pigments possessing a high refractive index.

The use of fillers results in an increase in density of the resin, the specific gravity of the combination tending to approach that of the filler material with increasing proportion of fillers—see *Fig. 4.1*. At the same time the inherent shrinkage of the resin on setting is reduced—see *Fig. 4.2*. The latter can be a particularly favourable characteristic in the case of making castings.

The effect on impact strength is

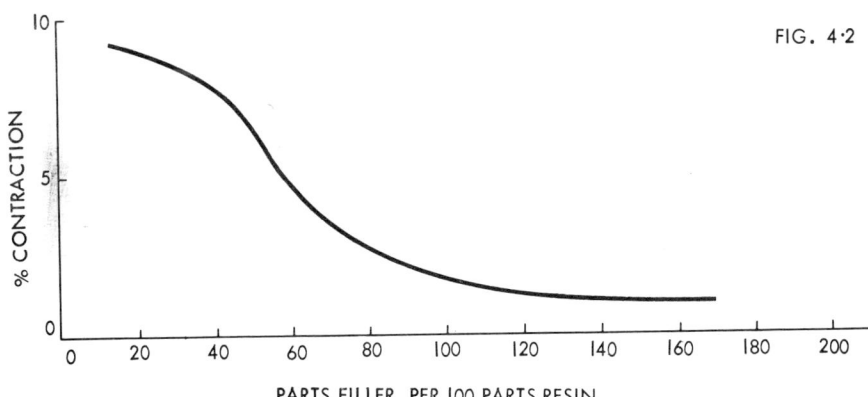

FIG. 4·2

SHRINKAGE RELATED TO PER CENT FILLER ADDED TO MIX

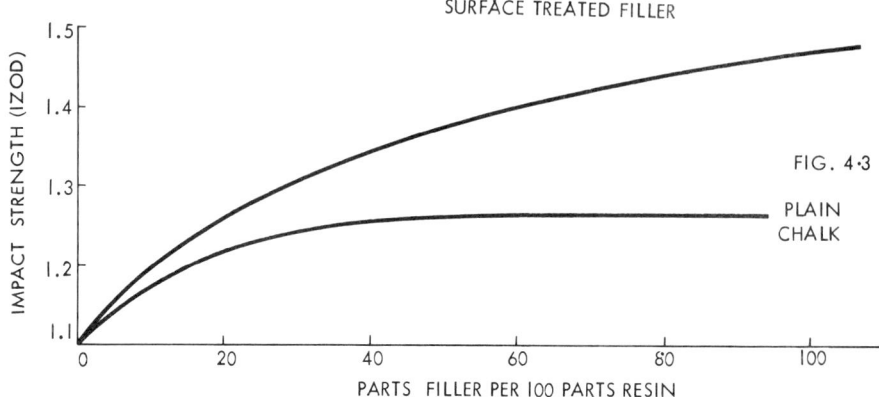

IMPROVEMENT IN IMPACT STRENGTH WITH ADDITION OF FILLER

shown in *Fig. 4.3*, where the superior performance of a surface-treated filler is obvious.

On the other hand, the addition of filler tends to increase the gel time, which may be a disadvantage; and, more important, an excess of fillers will tend to make the cured laminate brittle. For this reason, where fillers are used, it is generally recommended that the proportion of fillers to resin should not be in excess of 50:100. For general work, where fillers are used, the proportion is usually between 25 and 50 parts per 100 parts of resin, see Table 4.

Use of other types of fillers can be summarised briefly:

*Slate powder*—for general "bulking" of the resin, with the addition of a grey colour.

*Mica powder*—for casting and potting applications.

*Silica flour*—to improve abrasion resistance

*Aluminium powder*—for "metallic" mouldings

*Metallic flakes*—for opalescent metallic mouldings and colouring.

*Iron powder* or powdered silver—for moulding magnetic cores, etc.

*Graphite*—for moulding semi-conductors.

*Asbestos powder*—for self-extinguishing or flameproof mouldings

*Antimony oxide*—for self-extinguishing mouldings.

*Sawdust*—for producing resin/filler mixes that set with the appearance of solid wood.

*Vinyl flakes*—for "pearlescent" decorative effects.

Chalk and China clay are also used in very high proportions for making cold-setting "china castings". In this case only enough resin is used to provide a suitable bond for the powder. Similarly, metal powder fillers are used for producing "cold casting" metals on the same basis.

*Pigments*

Pigments may be used for through colouring (i.e. added to the main resin mix), or more usually these days, for colouring the gel coat only. As a general rule, no more pigment should

*Table 4*

Weight of Fillers Required

(Note: Usual proportions recommended are 25–50%)

(Weight in pounds, except where noted)

| Weight of Resin | Per cent Filler | | | | | | |
|---|---|---|---|---|---|---|---|
| | 20 | 25 | 50 | 75 | 100 | 125 | 150 |
| 10 lb | 2·0 | 2·5 | 5·0 | 7·5 | 10 | 12·5 | 15 |
| 5 lb | 1·0 | 1·25 | 2·5 | 3·75 | 5 | 6·25 | 7·5 |
| 2 lb | 6·4 oz | 8 oz | 1·0 | 1·5 | 2 | 2·5 | 3 |
| 1 lb | 3·2 oz | 4 oz | 8 oz | 12 oz | | 1·25 | 1·5 |
| 8 oz | 1·6 oz | 2 oz | 4 oz | 6 oz | 10 oz | 10 oz | 12 oz |
| 4 oz | 0·8 oz | 1 oz | 2 oz | 3 oz | 4 oz | 5 oz | 6 oz |
| 2 oz | 0·4 oz | ½ oz | 1 oz | 1½ oz | 2 oz | 2½ oz | 3 oz |

be added than is necessary to achieve the required degree of colour or opacity as most pigments tend to detract from, rather than enhance, the mechanical properties of the resin. Some types of conventional pigments may also have an inhibiting action on the resin and must be avoided. These include black pigments based on carbon.

Pigments are normally produced in the form of solids which must be ground into a fine paste with a suitable vehicle—in this case the resin. Ready-prepared pigments can be obtained in the form of resin-pastes, and this is the best way of using them, although it is not always possible to know whether the paste is highly concentrated or dilute. A maximum of 5% pigment should not be exceeded, which in the case of a highly concentrated paste usually means a maximum of 10% paste. With a diluted pigment paste, a 10% addition to the resin may not give the required depth of colour or opacity.

A wide variety of colours are available as pigment pastes, covering virtually all requirements. Specific types may also be suitable for inter-mixing to obtain different shades. The fastness of the colour depends primarily on whether the particular pigment used is fugitive or not, and also whether the pigment contains a proportion of dye. In the latter case the colour may change during curing because of the oxidising action of the catalyst having a bleaching effect.

The addition of pigments may also affect the gel time of the resin. In some cases the gel time may be increased (commonly with blacks and blues); in others it may be shortened. This can only be found out by experience with particular pigments.

If pigments other than specified as suitable for polyester resins are to be used, then the effect of these on the resin (and possible change of colour with the resin) should be investigated first by a few simple trials. This could avoid disappointing results on a finished moulding. Such trials are unnecessary on polyester resin pigment pastes.

A pigmented gel coat is the obvious, and the most satisfactory method of producing a coloured moulding, although this will only produce a high class finish where the moulding is produced in a polished mould. It is almost impossible to produce a satisfactory self-coloured finish on the "rough" side of a moulding where the surface has to be worked over to produce a good finish, particularly as a gloss colour will only show up the irregularity of the surface. For this reason the "rough" side of mouldings are best left uncoloured; or if colouring is thought necessary, sprayed with a matt finish (preferably with a flock or speckled effect).

It can be mentioned that despite the obvious advantages of a self-colour applied in the gel coat of a female moulding, some manufacturers of high class products still prefer to produce a "plain" moulding and finish by spray painting. This is found in the car industry, for example, with some moulded GRP body shells.

### Translucent Colours

Compatible dyes are also available for colouring polyester resin. These are intended for use where translucent through-colouring is required, such as in the production of rooflight mouldings and imitation stained glass, etc. These are generally referred to as translucent colours. Unless specially formulated for use with polyester resin, their possible effect on gel time and cold curing should be investigated by trials first.

### Metallic Colours

Metallic powders such as aluminium and bronze can be incorporated in the resin, but can be disappointing when used in the gel coat in a female mould. This is because a good metallic finish relies on the metallic particles aligning themselves in a preferred direction in a thin layer of "carrier" medium. They tend to become dispersed and random orientated in a thicker gel coat. Nevertheless, quite outstanding effects can be produced with vinyl flakes in thin gel coats, although the technique needed has to be developed by trial and error.

### Parting Agents (Release Agents)

One of the virtues of polyester resin is that it will adhere strongly to most surfaces. Equally, this can be a disadvantage when a lay-up has to be removed from its mould. To avoid the lay-up sticking to the mould it is necessary to prepare the mould surface initially with a non-stick surface. This can be done by covering the mould surface with a material to which polyester will not stick, e.g. cellophane or polythene; or, much more conveniently, by coating the mould with a release agent, which has the same effect of providing a non-stick skin on the mould surface.

The choice of a suitable release agent is influenced to a large extent by the size and shape of the mould, and the surface finish. The following are suitable types of release agents:

*Wax polish.* A good *carnauba wax* polish is a very effective release agent, and one of the most widely used on large, smoothly finished moulds. Most domestic and car-type hard wax polishes are based on carnauba wax, but may also contain other ingredients which can affect the setting of the gel coat, or even produce local adhesion. Silicone waxes should be avoided. In case of doubt, a wax polish should be tested for "release" and "adhesion" before being accepted as suitable. As a precaution, wax polishing followed by a final coating of a second release agent (e.g. PVA) can be used.

*Wax emulsions.* These have a similar action to wax polishes but can be applied more easily since they are in liquid form. They are then polished after drying. As a general rule they should then be coated with a second release agent (e.g. PVA).

*Table 5*

Release Agents

| Type | Method of Application | Suitable types of moulds | Removal | Remarks |
|---|---|---|---|---|
| Hard Wax (Wax Polishes) | Apply generously then polish | GRP, plastic, hardboard, wood, etc. | Xylene, or scrub with strong detergent | Silicone-waxes should be avoided. Additional protection given by further coat of PVA |
| Wax Emulsions | Smear on, leave to dry, then polish | Plaster, GRP, plastic hardboard, wood, etc. | Xylene or scrub with strong detergent | Suitable for one-coat treatment on all kinds of moulds |
| Candle Wax | Smear on | Small moulds and undercuts | Warm to melt | Difficult to apply evenly |
| PVA | Spray, brush or sponge (latter often easiest) | GRP, and all moulds with hard, sealed surface | Wash with warm water, or strip off if possible | |
| Cellulose Acetate | Spray preferred | GRP, plastic, hardboard, wood, etc. | Cellulose thinners | Final wax polishing recommended |

*Polyvinyl alcohol (PVA or PVAL)*

This is available as a concentrated liquid for dilution, or as a solution in water or a suitable solvent. It is available in both colourless and coloured forms. It is easy to apply since it can be sprayed, brushed or sponged on to the surface of the mould. One disadvantage, however, is that PVA has a low viscosity so it tends to drain off vertical surfaces, and can also accumulate in sharp corners, where it will take a long time to dry. PVA can be used on its own over small, non-porous mould surfaces; but on larger moulds it is more usual to use it as a second line of defence in conjunction with wax polishing.

PVA is also available in powder form, which can be dissolved in hot water to produce a suitable solution. The addition of a little glycerine is recommended to render the film slightly flexible, so that it can be stripped off the mould in a continuous sheet. The addition of alcohol to the solution, or the use of alcohol instead of water as a solvent, produces a solution which dries much more rapidly.

*Cellulose acetate.* Cellulose acetate dissolved in acetone, i.e. cellulose "lacquer" is also an effective parting agent, particularly suitable for applying by spray. An additional coating of wax polish is, however, recommended for complete treatment. It should be noted that cellulose *nitrate* solutions are not suitable release agents, and the majority of cellulose lacquers are based on nitrate rather than acetate.

*Candle Wax.* This is an effective parting agent, but rather difficult to apply evenly except to small moulds. In certain cases, however, it may be useful for treating undercuts.

Main properties of these types of release agents are summarised in Table 5.

*Sheet Release Agents*

Where the mould lends itself to covering by draping with thin sheet material, the following can be used instead of applying parting agent direct to the surface of the mould.

*Acetate sheet.* Preferably in the form of cellophane (0·002–0·003 in. thick) of the non-waterproofed kind (which contains lacquer and will not part from the mould). Note again that cellulose nitrate sheet ("celluloid") is not effective.

*PVA sheet.* Very effective, and available under the trade name of "Pevalon".

*Polythene.* Effective, but subject to wrinkling.

*Neoprene.* Effective, but expensive and not readily available. It has the advantage of being a stretchable material.

*Technique*

The answer to obtaining a good, clean release from a mould does not lie so much with the parting agent used but with the care with which it is applied. Thorough application and careful inspection to ensure that the whole area is covered by the agent used will be rewarded with a clean release and a surface on the finished laminate which requires no work to restore it to standard. As the resin accurately follows every mark and scratch on the surface of the mould, and also shrinks slightly as it sets, it will be adhering very closely to the surface of the parting agent.

Whenever a release or parting agent is used on a moulding which is subsequently to be painted, it is essential that all traces of the release agent must be removed from the surface of the mould. In the case of certain waxes complete surface cleanliness is difficult to achieve, even with strong degreasing agents, and in special circumstances may call for a two-layer application of parting agent over the surface of the mould. The first layer, wax, then comprises the main agent, the second, polyvinyl alcohol, a coating which protects the moulding from the wax release agent.

# 5.  MOULDS

GRP itself cannot be formed to shape. In its "wet" state it is far too floppy and sticky to manipulate. In its cured state it is rigid and unworkable, except by machining. All shapes in GRP have, therefore, to be laid up in moulds, which duplicate the contours of the finished shape required. The mould itself may be of *female* or *male* form (*Fig. 5.1*.)

Choice of the type of mould depends on which is to be the smooth side of the finished moulding, this always being the side laid up in contact with the mould. Thus shapes like car body panels, boat hulls, and so on would normally be laid up in *female* moulds to produce a smooth surface on the outer side. Shapes like baths, basins, etc., would be laid up on a *male* mould, so that again the side that shows on the finished job is smooth. In other cases it may not matter which side is rough and which is smooth, when the choice of mould type can be based on which is the simplest to make, or which would be the best side for the smooth surface judged

from a functional rather than a decorative point of view.

Moulds can be constructed from almost any material—plasticine, clay, plaster, wood, hardboard, ply, sheet metal. However, best results with female moulds are undoubtedly obtained by making the mould itself in GRP, which means a two-stage process of construction. First a full size pattern is made (usually called a plug), and suitably finished. GRP is then laid up on this, producing a moulding which is a faithful reverse pattern of the plug. When set, this is removed and forms the mould from which subsequent mouldings can be made. Obviously such a mould can be used over and over again for repetitive production. It is a relatively costly, and lengthy, method of making a mould for one-off, but it is still the best one where a female mould has to be used.

From the point of view of cost and time saving, it is better to use a male mould for one-off jobs. The full size pattern can then be used as the mould. The same technique can be applied to

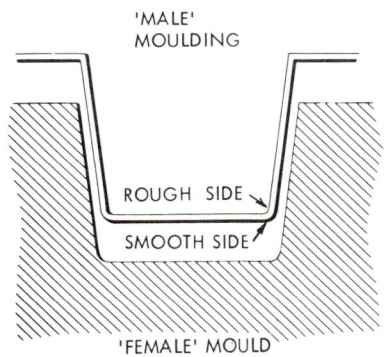

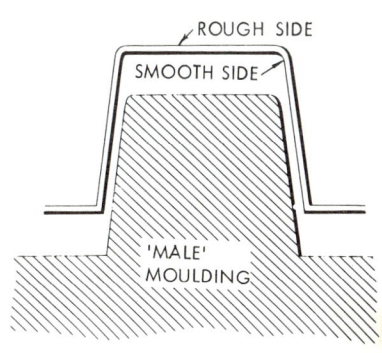

FIG. 5·1 THE TWO BASIC TYPES OF MOULDS

a female mould, but unless restricted to simple shapes and forms, the making and finishing of the pattern can become quite difficult. Remember that any pattern made for direct use as a mould must be a reverse image of the final shape required. Cost may be the deciding factor. For small objects the making of a female mould in GRP for one-off jobs is usually more than justified, because of the simple plug construction that can be used. With a large object, the more difficult job of making a reverse image pattern to act as a female mould can be worthwhile, because of the considerable saving in glass fibre and resin otherwise involved in making a GRP mould.

Do not overlook the possibilities of using an existing article as a pattern for a mould, or as a basic shape which can be built upon to produce a pattern or plug. This can often save considerable time and effort in patternmaking.

Full size patterns, whether for use as plugs or direct moulds, have one thing in common—they must be finished to a perfectly smooth surface. Also, of course, they must duplicate the shape required faithfully. The materials used can be selected on the basis of providing the simplest, cheapest and quickest method of construction. In some cases an existing object can be used directly as a plug, provided it is of suitable shape and the surface is, or can be, finished glass smooth.

Wood is an obvious choice for making small patterns or plugs, balsa wood being particularly recommended because of the ease with which blocks and sheet can be cemented together to produce a basic form and the ease with which balsa can be carved to shape. The main difficulty with this material is the considerable time and effort needed to seal the grain and obtain a final glass-smooth surface. The cost of balsa also excludes its use for larger jobs.

Here the cheapest type of pattern or plug construction is undoubtedly plaster, unless the form required can be built up simply from sheet hard-

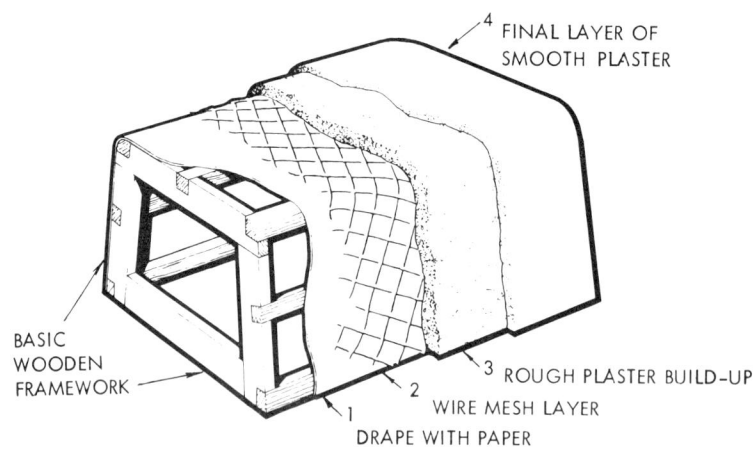

FIG. 5·2   SIMPLE PLASTER PATTERN CONSTRUCTION
            FOR LARGE SHAPES

board. A plaster pattern can be built up around a rough mock-up of the shape, to save time, weight and the amount of plaster used. An example of a simple mock-up is shown in *Fig. 5.2*. This is then draped with brown paper, followed by a layer of wire mesh on which the plaster is trowelled and built up to a suitable thickness. The wire mesh provides a base on which to lay the plaster and the paper backing prevents it falling through whilst still wet. On larger patterns it may be necessary to support the plaster by draping the wire mesh with a further layer of hessian, or hessian soaked in plaster. In this case a wider mesh can be used see *Fig. 5.3*.

The final coat of plaster should be trowelled on and smoothed out as far as possible. When set, it must be worked over to produce a uniform, smooth surface. The surface will be quite porous, so will require sealing before it can be used as a plug (or a mould). Cellulose fillers are particularly recommended for sealing plaster for they provide a good surface which can be further rubbed down with fine wet or dry abrasive. Alternatively, several coats of shellac or two or three coats of any two-part synthetic resin sealer or varnish can be used.

Composite construction is also a possibility. A basic shape may be built up in "box" form from hardboard, and the shape further extended by plasticine or simple built-up constructions. The whole could then be plastered over (or parts only plastered

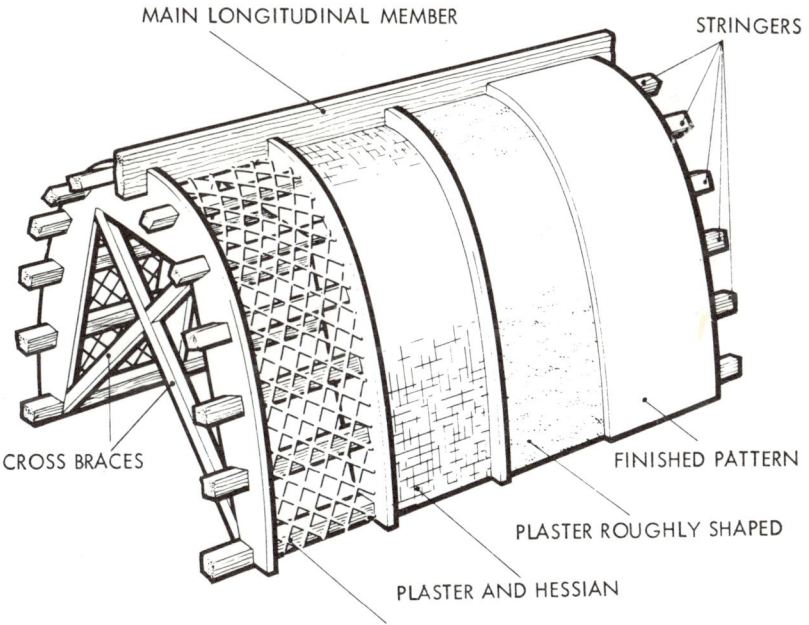

FIG. 5·3 MORE ELABORATE PLASTER PATTERN

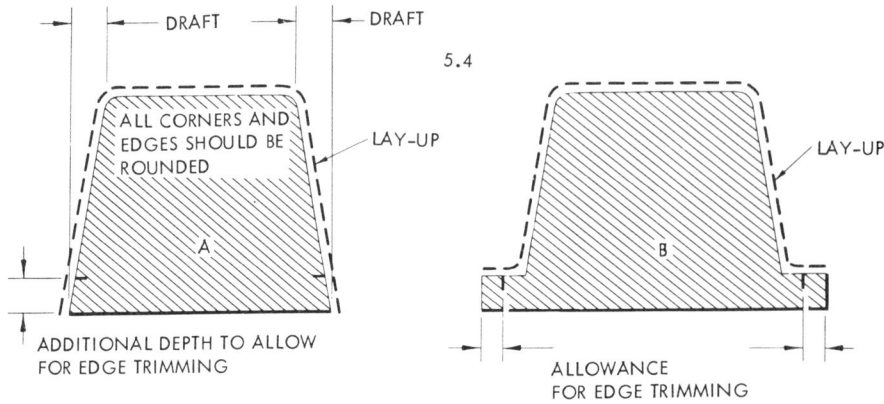

5.4

(A)      PATTERN FOR MOULDING WITH PLAIN EDGE

(B)      PATTERN FOR MOULDING WITH ROUNDED EDGE

over) to yield a uniform, rigid surface which can be sanded down quite smooth.

Regardless of the materials used for the construction of a plug, or a direct mould, this question of finishing glass-smooth with a fully sealed surface is all important. A GRP lay-up will faithfully duplicate any defects in the surface. Porosity of the surface will encourage penetration of resin, even if apparently protected by the parting agent applied. At best, this will produce a rough surface on the moulding in this area. At worse, the moulding will stick.

The shape of the pattern must also take into account the fact that any GRP moulding taken off it will be rigid. Thus a male pattern should have draft or taper on the vertical surfaces so that the moulding does not become locked in place—*Fig. 5.4*. It should also be of sufficient extra size to allow for the fact that the edges of a GRP lay-up will be starved of resin, and relatively weak. They must therefore come outside the final shape. Similar considerations apply in the case of a female pattern. Where the pattern is used as a plug, of course, a male pattern produces a female mould; and a female pattern a male mould.

If it is strictly necessary to accommodate reverse tapers or undercuts in the pattern, then the mould derived must be split. Some examples are shown in *Fig. 5.5*. Split moulds are to be avoided, whenever possible, especially in amateur construction.

Where the pattern is to be used to produce a GRP mould, provision should be made on the plug to ensure that the edges of the mould are flanged. This will add considerable stiffness to the mould. It is well worthwhile to bring the plug up to the highest finish possible, so that the finish obtained on the surface of the GRP mould will be as near perfect as possible. Whilst surface defects on the GRP mould can be smoothed and polished out, this is a lengthy and tedious process. It is much better to eliminate the necessity for such re-working by putting the extra time necessary into finishing the plug to achieve perfection of finish.

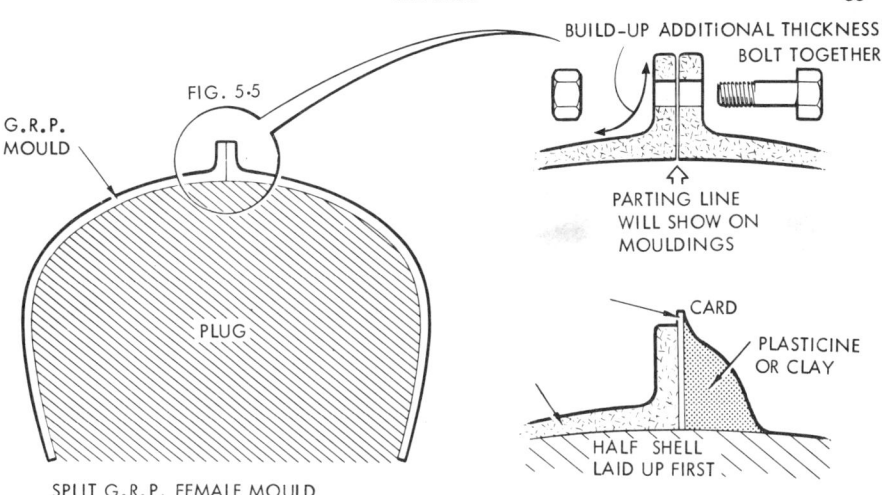

FIG. 5.5

G.R.P. MOULD

PLUG

SPLIT G.R.P. FEMALE MOULD

BUILD-UP ADDITIONAL THICKNESS
BOLT TOGETHER

PARTING LINE
WILL SHOW ON
MOULDINGS

CARD
PLASTICINE
OR CLAY

HALF SHELL
LAID UP FIRST

A GRP mould is laid up on a plug in a similar manner to any GRP construction—see Chapter 7. About the only difference is that a slightly thicker gel coat should be used—0·020–0·025 in. thick, or approximately 2 ounces per resin per sq. ft. A fast-setting resin should be used for preference. Moulds with large vertical surfaces should use a thixotropic resin to prevent draining. Once the gel coat has set, a layer of surface tissue should be applied with resin, followed by succeeding layers of glass fibre mat to build up to a suitable thickness.

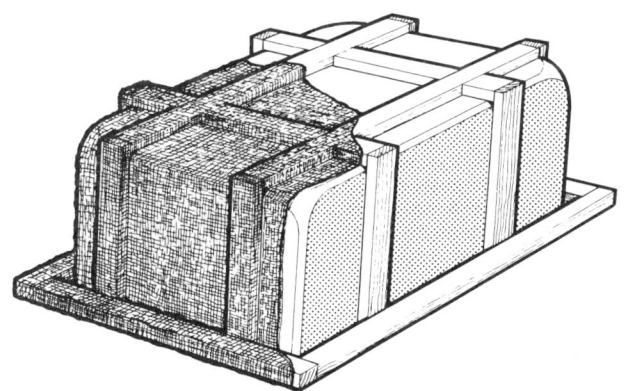

FIG. 5·6    EXTERNAL STIFFENING APPLIED
TO A FEMALE G.R.P. MOULD

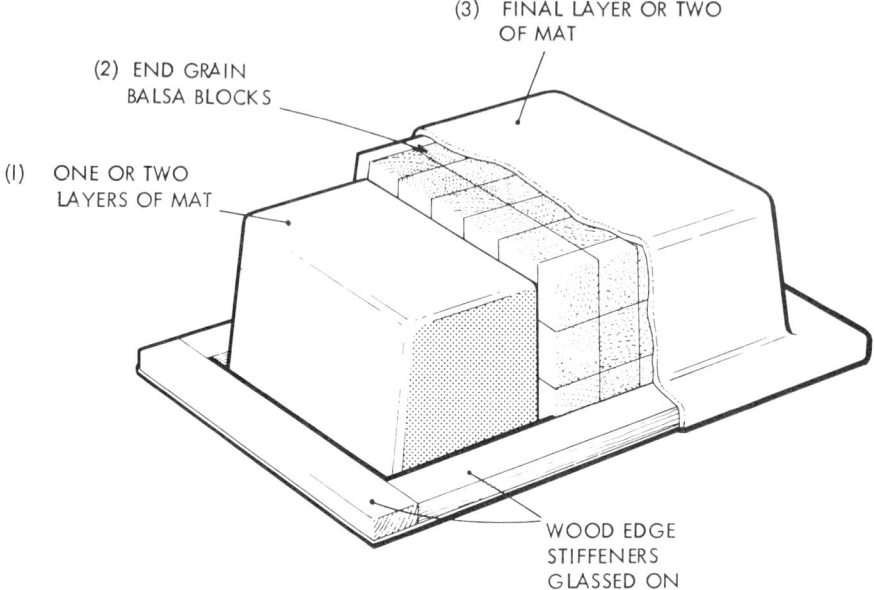

(3)  FINAL LAYER OR TWO
OF MAT

(2)  END GRAIN
BALSA BLOCKS

(1)  ONE OR TWO
LAYERS OF MAT

WOOD EDGE
STIFFENERS
GLASSED ON

FIG. 5-7    FEMALE G.R.P. MOULD WITH BUILT-IN STIFFENERS
(SANDWICH CONSTRUCTION)

The thickness required obviously depends on the size of the mould. As a rough rule, work to a thickness of $1\frac{1}{2}$ times that which would be considered suitable for the actual *moulding* (to be taken off the mould). It need not be any thicker than that for extra stiffness and rigidity can be given by adding bracing strips, etc., to the outside of the mould, in the manner shown in *Fig. 5.6*. A further way of stiffening a GRP mould without the expense of additional mat layers or the addition of external stiffeners is to use balsa blocks in "sandwich" construction—*Fig. 5.7*.

It is very important that a GRP mould should be rigid enough to resist distortion or bending when removed from the plug. It must be absolutely "solid", so that it maintains the shape of the plug faithfully when removed from it, and when being worked on in using it as a mould to produce a lay-up moulding. This means anticipating the amount of external reinforcement necessary whilst the GRP mould is still on the plug. All such reinforcement should be added, and the mould *left* on the plug, *until fully cured and hardened*. In the case of a moulding where the shape is critical, this means a minimum of two days, unless post-curing can be applied.

This chapter has treated the subject of moulds generally. Moulds are essentially individual to particular jobs, and so further information on this subject will be found under the separate headings dealing with specific constructions.

# 6. TOOLS AND EQUIPMENT

Working with glass fibre is different from handling any other material. Glass fibre cloth or mat is floppy and difficult to cut neatly. Once impregnated with resin it is even more difficult to work neatly, and messy. Wear old clothes, or a suitable protective overall or apron, and protect the working area from surplus resin which can stick and harden on it, and be very difficult to remove. Hands can be protected by using a barrier cream (e.g. Kerodex or Rozalex); or by wearing thin rubber gloves (disposable polythene gloves are particularly recommended as being cheap and easy to work in).

Glass fibre mat can usually be torn to shape, although this should only be attempted with dry hands. Cloth, tissue and tape needs cutting to shape. Here scissors can be used, or a really sharp knife or razor blade, although cutting edges will rapidly become dulled.

Some form of weighing machine is needed to measure out the required quantities of resin, together with suitable measures for arriving at the correct proportion of catalyst to be used. Alternatively, liquid measures can be used throughout (see Chapter 3).

Other tools and equipment required

FIG. 6·1    BRUSHES

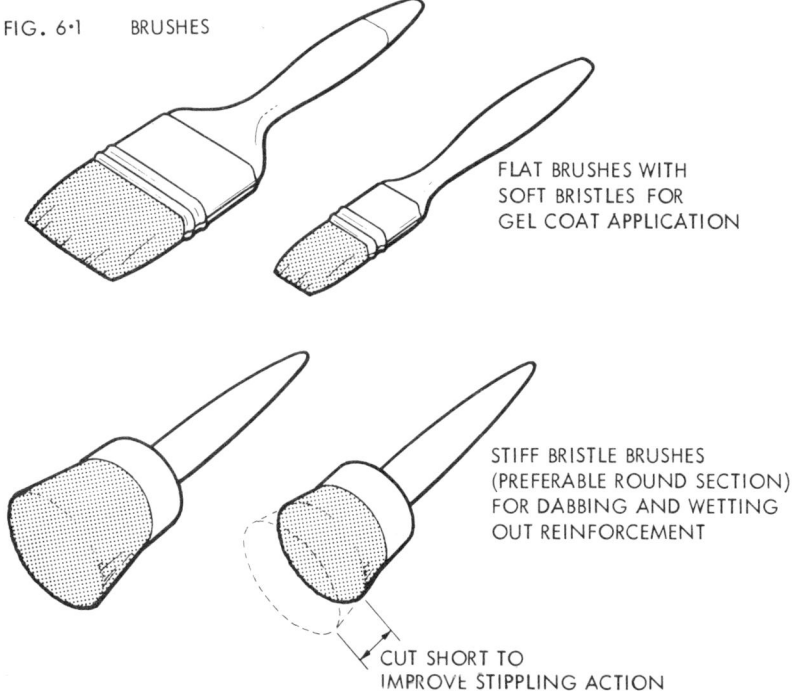

FLAT BRUSHES WITH
SOFT BRISTLES FOR
GEL COAT APPLICATION

STIFF BRISTLE BRUSHES
(PREFERABLE ROUND SECTION)
FOR DABBING AND WETTING
OUT REINFORCEMENT

CUT SHORT TO
IMPROVE STIPPLING ACTION

FIG. 6·2    ROLLERS FOR HAND LAY-UP OR ART MOULDING

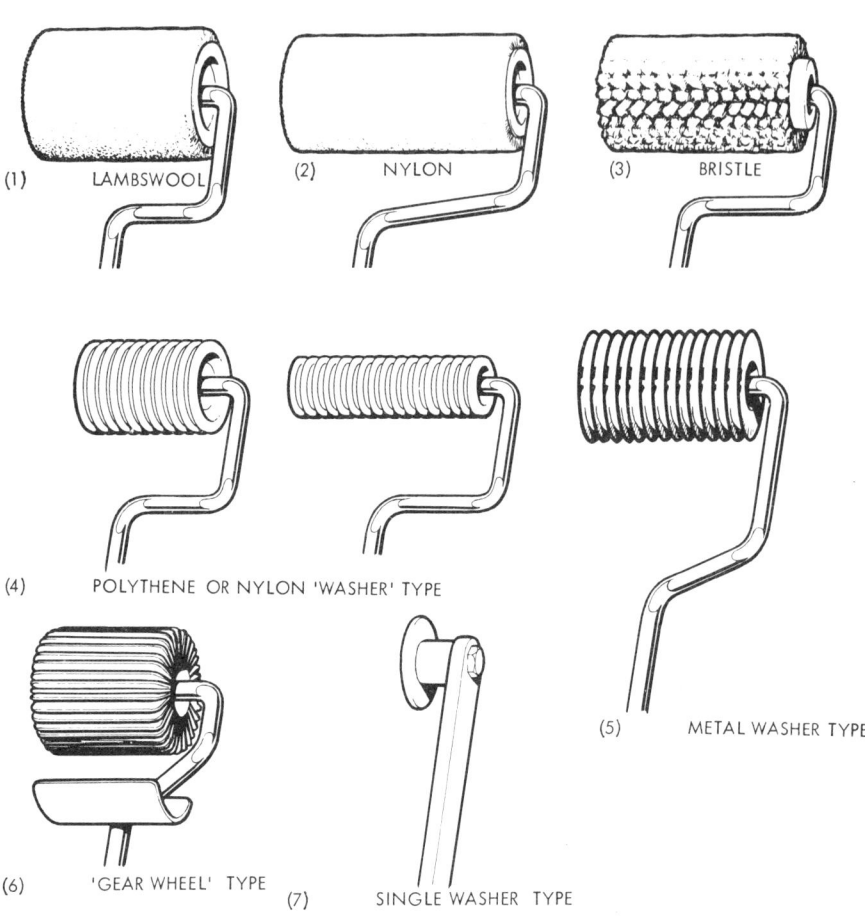

(1)    LAMBSWOOL

(2)    NYLON

(3)    BRISTLE

(4)    POLYTHENE OR NYLON 'WASHER' TYPE

(5)    METAL WASHER TYPE

(6)    'GEAR WHEEL' TYPE

(7)    SINGLE WASHER TYPE

are then described under the separate headings which follow.

*Brushes (see Fig. 6.1)*

Brushes are used to *paint* on the gel coat, and "*stipple*" or dab the glass fibre reinforcement in place to ensure uniform wetting with resin.

Standard paintbrushes are suitable for gel coat application, although the preference is for ones with white or colourless bristles rather than black bristles, so that loose bristles will not show up. Sizes used depend on the size of the mould. For large areas it will be more convenient to use 2 in. or

3 in. brushes; although 1 in. and 1½ in. sizes will be more suitable for general work.

The same brush can be used for *applying* the resin during the lay-up. To dab and stipple the glass fibre in place, however, a stiff circular brush is to be preferred, if necessary cutting the bristles short to get a good rigid brush. Final consolidation can then be done with a roller, if the job is large enough.

*Cleaning Solutions*

Brushes should be cleaned immediately after use in polyester solvent cleaner, acetone or cellulose thinners and can be left standing in such solvents for a time, if necessary. Final cleaning should then be done by washing out in strong detergent solution and then allowing to dry. They should never be left standing in water (or solvent) indefinitely. When used, brushes should always be clean *and dry*.

*Rollers (see Fig. 6.2)*

Rollers are used instead of brushes for consolidating the glass reinforcement in the lay-up where large areas are involved. They produce a more uniform "squeegeeing" effect, and the job can be done much faster. Lambswool paint rollers can also be used for applying the gel coat on large, flat areas but although this speeds the work the resulting thickness of resin is usually inadequate from a single coating. The job therefore has to be done at least twice, allowing each coat to gel before applying the next.

Ordinary paint rollers can be used, although they can be tedious to clean. Special types include rollers with bristles, and those with serrated or washer configuration. The latter are usually made in metal, but sometimes plastic (e.g. polythene or nylon). Choice is largely a matter of personal preference, although most professional work is now done with "hard" rollers. There is, however, some advantage in using a "soft" roller for certain jobs. Note, too, that several sizes of rollers may be useful—long rollers for working over large surfaces and short rollers for working in restricted corners, etc. Alternatively, a stiff brush can be used in such places.

Rollers can be cleaned in the same solution as brushes, although metal rollers are often more conveniently cleaned by burning off the resin and then scraping clean.

FIG. 6·3    KNIVES

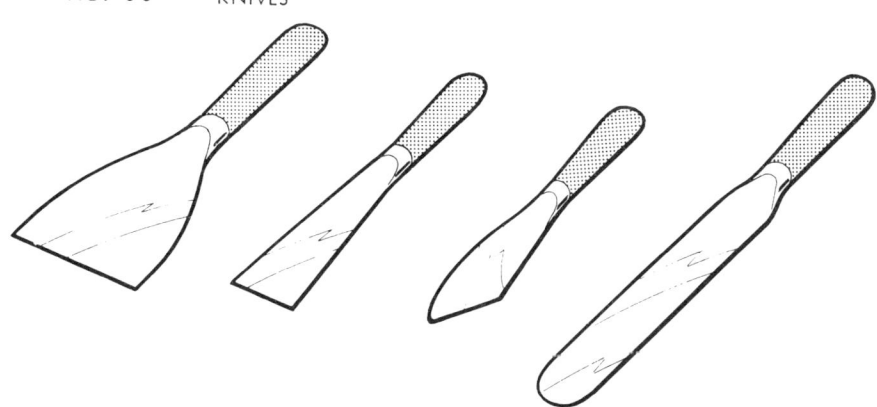

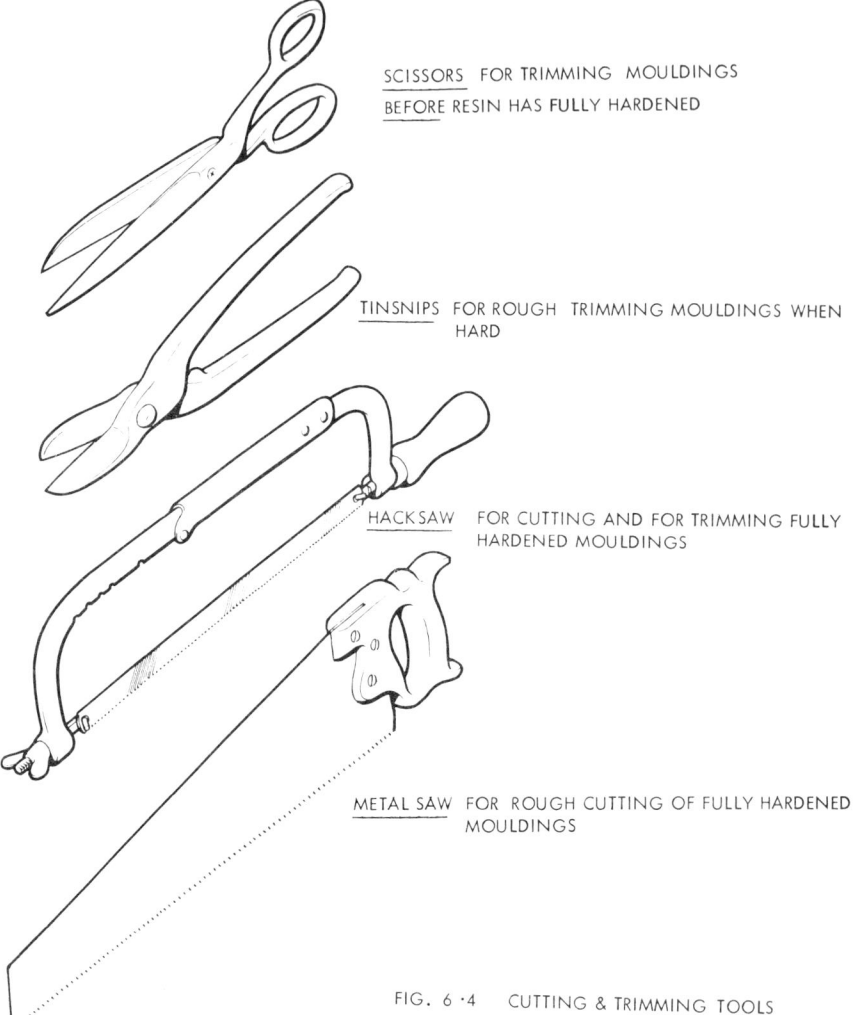

SCISSORS FOR TRIMMING MOULDINGS BEFORE RESIN HAS FULLY HARDENED

TINSNIPS FOR ROUGH TRIMMING MOULDINGS WHEN HARD

HACKSAW FOR CUTTING AND FOR TRIMMING FULLY HARDENED MOULDINGS

METAL SAW FOR ROUGH CUTTING OF FULLY HARDENED MOULDINGS

FIG. 6·4    CUTTING & TRIMMING TOOLS

### Knives (see Fig. 6.3)

Palette knives are useful for working pigments and fillers into the resin, and for small stopping or filling jobs using resin-filler. Larger jobs of this nature can be tackled best with conventional putty knives.

### Cutting Tools (see Fig. 6.4)

Once hardened off, a GRP moulding will be too hard to be cut or trimmed with woodworking tools, and getting a clean cut edge on the moulding can be hard work. Either a metal-cutting saw or a hacksaw must be used. The latter makes the neatest job, using a blade with 32 teeth to the inch. The blade should be discarded as soon as it shows signs of becoming blunted.

Note here that much effort can often be saved by trimming the edge of a moulding during the hardening time, i.e. before the GRP has reached its full hardened strength. At this stage trimming can still be done with a sharp knife, or possibly even scissors. At a slightly later stage, but before the GRP has set really hard, trimming can be done with tinsnips.

### Sanding Discs

These should be of "hard" type with a coarse grit on a tough backing. Fine abrasive discs will clog too rapidly to be effective. Sanding discs used on power drills are the quickest and easiest way of final trimming edges, or removing rough spots or irregularities on the surface of finished laminates.

A sanding disc, or orbital sander, is also effective for flatting down resin-filler when set. If the build-up is excessive, a "Surform" tool with a new blade should be used first to level off as far as possible. This will be quicker than using a sanding disc, and will also greatly increase the life of the disc. A "Surform" tool is also useful for rough shaping and trimming all GRP mouldings.

### Files

To cut effectively, files should be of coarse metal-cutting type, and new. The life of a file used on glass fibre is very limited, and it should be discarded when it shows signs of becoming blunt.

### Drills

Glass fibre must be considered as metal for the purpose of drilling. Only high speed steel twist drills should be used, with the lowest possible drilling speed (particularly with larger drill sizes). Drilling should always be done from the gel coat side, and the other side of the moulding backed up with a block of wood to limit the extent of tearing or cracking when the drill breaks through. A hole drilled from the opposite side will almost invariably damage the gel coat, even if backed up.

Drills used for drilling glass fibre should always be sharp, and ground to the same point angle as for cutting metals. In the case of drill sizes of $\frac{3}{8}$ in. and above, the point angle of the drill can be reduced with some advantage. No lubricant is normally required, but if a drill does show signs of running hot, water can be used to cool it.

### Abrasive Papers

Any "finish sanding" on GRP mouldings should always be done with wet and dry abrasive papers, used wet and frequently rinsed off to clean. Some authorities also recommend the addition of a little soap to the rinsing water to act as a lubricant.

For "rough" sanding 240 grit will be suitable. Progressively finer grades down to 500 or even 600 grit can be used for final smoothing. The surface can be further improved by buffing and polishing with metal polish, polishing compound or jeweller's rouge. A final overall finish can be given by wax polishing, provided this is acceptable to the application.

# 7. BASIC TECHNIQUE

THE same basic technique applies in all hand lay-up or "contact" mouldings —so-called because the resin/glass fibre is laid up in contact with the mould, without the application of pressure, other than that used to consolidate the glass fibre reinforcement and ensure that it is thoroughly wetted. The stages involved are described, in order, under separate headings.

## Choice of Materials

General purpose *laminating resin* is used for the majority of applications, together with appropriate catalyst (hardener), and resin accelerator (unless the resin is of pre-accelerated formulation). Other types of resins and resin additives are described in Chapter 3.

*Glass fibre mat* is normally used for reinforcement, the popular choice being 1½ ounce weight. Glass fibre tissue or surfacing tissue may also be employed. Other types of glass fibre reinforcements are described in Chapter 2.

## Quantities Required

The quantity of glass fibre mat (or woven cloth) required can be worked out on an area basis. Work out, or estimate, the surface area of the mould and add 5% to give the area of glass fibre required for a single layer to be on the safe side and allow for edge trimming. The total area of glass fibre required can then be found by multiplying by the number of layers it is intended to use. Note, however, that this figure may need adjustment since certain mouldings may be laid up with different number of layers in different areas.

In the case of surface tissue, if used, the quantity is estimated as for a single layer.

The amount of *resin* required can be estimated on the basis of the following general figures:

*Gel coat*—estimate on the basis of 2 ounces of resin per square foot of mould surface area.

*Laminating*—estimate on the basis of a 30:70 glass:resin ratio for glass fibre mat (or surface tissue); and a 50:50 glass:resin ratio when using glass fibre cloth. In other words, work out the total weight of glass fibre to be used, and estimate the corresponding weight of resin required from the appropriate ratio—see Table 6.

Note that this is only a general figure, suited to average workmanship. It is possible to reduce the amount of resin and obtain a higher glass:resin ratio, which is favourable to strength, if particular care is taken to ensure that each glass layer is thoroughly wetted with resin during the lay-up. Equally, it is possible to use more resin, which will make it easier to ensure complete wetting out, but reduces the glass:resin ratio and the strength of the moulding. In the case of glass cloth reinforcement, which is more difficult to wet out, actual resin:glass ratios may be even more variable, depending on the type of cloth used, and individual technique.

## Mixing the Resin

The accelerator is commonly mixed with the resin, as supplied (pre-accelerated resin). If not, it should be added to the resin first in the recommended proportion.

If fillers or pigments are to be used, these should be added next and thoroughly dispersed by complete mixing, but avoiding excessive agitation which could introduce air bubbles into the resin. If necessary, the resin

*Table 6*

Amount of Resin required

(weight in ounces per sq.ft.†)

| Glass weight, ounces per sq. ft* | GEL coat | Glass mat reinforcement | Glass cloth reinforcement | Surface tissue |
|---|---|---|---|---|
| 1 | | $2\frac{1}{3}$ | 1 | |
| $1\frac{1}{2}$ | | $3\frac{1}{2}$ | $1\frac{1}{2}$ | |
| 2 | | $4\frac{2}{3}$ | 2 | |
| 3 | 2 ounces per sq. ft. in all cases | 7 | 3 | No extra allowance needed |
| $4\frac{1}{2}$ | | $10\frac{1}{2}$ | $4\frac{1}{2}$ | |
| 6 | | 14 | 6 | |
| $7\frac{1}{2}$ | | $17\frac{1}{2}$ | $7\frac{1}{2}$ | |
| 9 | | 21 | 9 | |
| $10\frac{1}{2}$ | | $24\frac{1}{2}$ | $10\frac{1}{2}$ | |
| 12 | | 28 | 12 | |

\* Multiply mat or cloth weight per square foot by number of layers to find glass weight per sq. ft. E.g. 4 layers of $1\frac{1}{2}$ ounce mat would be equal to $4 \times 1\frac{1}{2} = 6$ ounces glass weigh per sq. ft.

† Total amount of resin required follows from amount required per square foot for type of reinforcement and number of layers as determined above, multiplied by the surface area of the moulding. E.g. 4 layers of $1\frac{1}{2}$ ounce mat laid in a mould of 4 sq. ft. surface area would require $14 \times 4 = 56$ ounces of resin.

should be left to stand until clear of air bubbles.

The exception to this rule is where an absorbent filler is used, e.g. sawdust. In this case the resin should be activated with catalyst before the filler is added and mixed in. This will ensure proper distribution of the catalyst through the resin. If the catalyst is added last, the filler will show a preferential absorption for the catalyst, which may starve the resin of catalyst and result in an incomplete cure.

The catalyst (hardener) should only be added immediately before the resin is required for use. Once added, the resin will have a strictly limited pot life. The proportion of catalyst is not critical, but use recommended quantities. Lacking any specific information on the subject, a 3% proportion of catalyst (by weight) can be considered suitable for general use. This may be increased to 4% to produce a faster gelling time, e.g. if the temperature is low; or decreased to 2% to reduce the gelling time, if the temperature is low. If the temperature is very low, then additional catalyst or "winter additive"

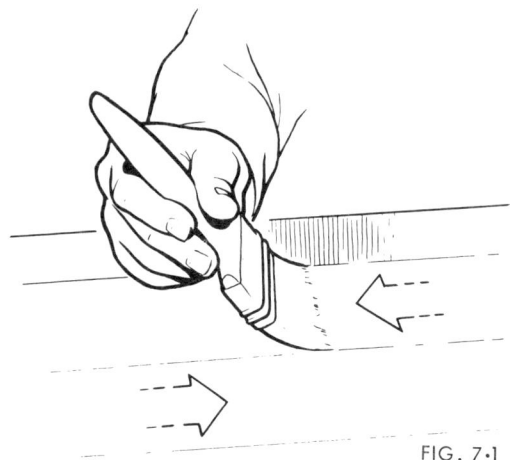

FIG. 7·1

PAINT ON GEL COAT WITH CONTINUOUS MOTION
(DO NOT PAINT BACKWARDS AND FORWARDS)

should be used, in proportions specified by the supplier.

*Working conditions*

Ideally the air temperature should be at least 50°F. If the temperature is lower than this, 4 % catalyst (hardener) should be used, or "winter additive". If the temperature is below about 40°F however, it is best to leave laying-up until conditions improve, or the area can be heated. "Winter additives" can promote gelling and setting in cold air, but the resulting moulding may suffer as a consequence because of incomplete curing.

More important still—the air should be dry, or reasonably dry. Never attempt laying-up in damp or excessively humid conditions as the moisture in the air will inhibit setting of the resin. This cannot be compensated for by chemical additives.

*Preparing the Mould*

The mould must be clean (removing any remaining parting agent if the mould has been used before), free from surface damage, and dust-free.

If necessary, it should be polished shortly before use, followed by treatment with parting agent (release agent).

*The Gel Coat*

If a self-coloured moulding is required, this is normally added to the gel coat only. The gel coat resin is thus prepared by mixing with a suitable pigment paste, avoiding the formation of air bubbles. Any pigment added must be dispersed thoroughly in the resin.

Note: if identical colour matching is required on two or more mouldings, enough coloured gel coat resin to do *all* the mouldings involved should be made up at the same time. Separate quantities of this basic colour mix can then be used, as required.

After activating with catalyst (hardener) the resin should be brushed in place, using a wide, soft brush and single, sweeping strokes. The aim should be to get an even distribution of resin in one continuous application, not brushing backwards and

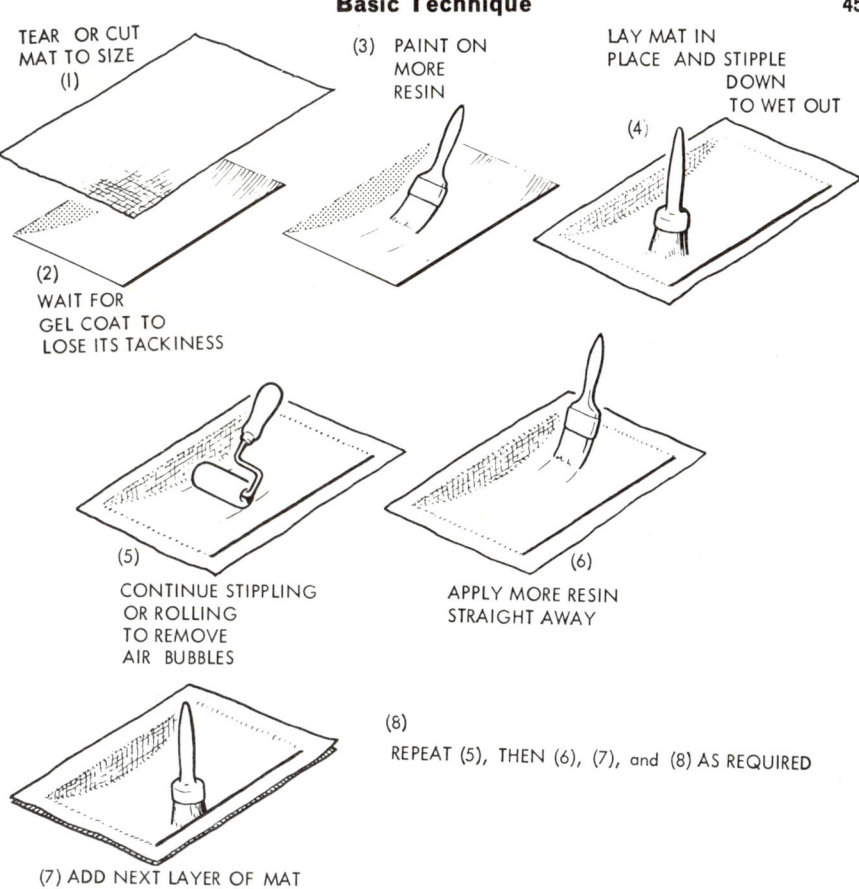

TEAR OR CUT MAT TO SIZE (1)

(2) WAIT FOR GEL COAT TO LOSE ITS TACKINESS

(3) PAINT ON MORE RESIN

LAY MAT IN PLACE AND STIPPLE DOWN TO WET OUT (4)

(5) CONTINUE STIPPLING OR ROLLING TO REMOVE AIR BUBBLES

(6) APPLY MORE RESIN STRAIGHT AWAY

(7) ADD NEXT LAYER OF MAT AND STIPPLE DOWN IN PLACE

(8) REPEAT (5), THEN (6), (7), and (8) AS REQUIRED

FIG. 7·2 WET LAY-UP

forwards (*Fig. 7.1*). In the case of near horizontal surfaces, the resin can often be poured in place and distributed with a brush. The aim should be to achieve a uniform gel coat thickness of between 0.01 and 0.02 in. If a lambswool paint roller is used to distribute the gel coat over large surfaces, it will be impossible to produce this thickness in a single coat. In that case the first coat should be left to gel, and second coat then applied. No attempt should be made to paint or roll over a first coat of resin until it has gelled or set to a condition where it is no longer sticky to the touch. Equally, avoid touching a gel coat which has not gelled, as this will fingerprint it. Test with the finger on an area which will subsequently be cut off the moulding when trimming to final shape.

*The Reinforcing Layers (see Fig. 7.2)*

No further work can be done in the mould until the gel coat has gelled, or set to the point where it is like a fairly rigid jelly with no surface tackiness.

A further batch of resin is then activated. This is normally clear resin, although pigment may be added for "through" colour if thought desirable. Fillers should *not* be used in the lay-up resin.

A coating of resin is painted over the gel coat, followed by laying the glass fibre in place. The first layer may be of glass tissue, to serve as a "screen" to mask the glass pattern in subsequent layers, but the use of surfacing tissue is largely a matter of personal preference. If not used, proceed directly with the first layer of mat (or cloth).

This should be cut or torn to suitable size, laid in place over the wet resin and stippled down in place with a stiff brush, using a *dabbing* motion only. Avoid sweeping or "painting" strokes, as this will only pull up the glass fibres, particularly in the case of mat. Dabbing down can be followed by rolling to ensure proper consolidation and wetting of the glass. The two important things to ensure are:

(i) All the glass is thoroughly wetted with resin. This will be shown by the change of colour of the glass from whitish to a translucent look. Make sure that no white patches remain.

(ii) The glass reinforcement is properly tamped down with no *air bubbles* trapped in or under it. It is *very important* that any air bubbles are removed as these will otherwise form weak spots in the moulding.

Glass fibre mat will mould readily to curved shapes, although it may prove difficult to bed down properly over sharper curves. It may be best in such areas to shred the mat before laying in place, to provide looser fibres which can be bedded down.

Opinion tends to differ on the matter of joints, where one piece of glass fibre has to carry on from another. To ensure maximum joint strength it is best to overlap the next piece over the first by about $1\frac{1}{2}$ in. to 2 in., and then work over this overlap to spread the two layers out to a uniform thickness (*Fig. 7.3*). Again this is easier if the second edge is shredded slightly first. Somewhat neater results can be produced by laying the second piece of glass fibre edge to edge and working the glass fibres together to intermesh. This is a more tricky operation, and there is a danger of thinning the reinforcement along the joint line. As a consequence the moulding will be weaker at this point because of the higher resin content. This can be

FIG. 7·3　BASIC JOINT TECHNIQUE

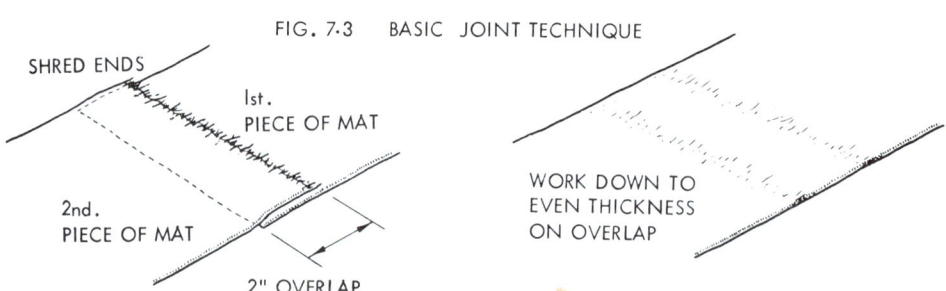

SHRED ENDS

1st.
PIECE OF MAT

2nd.
PIECE OF MAT

2" OVERLAP

WORK DOWN TO
EVEN THICKNESS
ON OVERLAP

offset to a large extent by making sure that joints in subsequent layers occur in different places.

Succeeding layers of glass reinforcement can be laid up, in turn, directly on the first whilst the resin is still wet, i.e. there is no need to let the first layer gel before applying the next. In fact, it is better to work through all the layers continuously, as this will produce the most homogeneous moulding.

The exception is where a large number of layers are being used. Four layers of $1\frac{1}{2}$ ounce mat (or more specifically, 6 ounces of glass reinforcement per square foot) is about the maximum thickness which can be laid up directly, one on top of the other, without the moulding becoming excessively hot. With thick mouldings, therefore, lamination should be stopped at this stage until the first layers have gelled and hardened off. Further build-up can then be attempted without fear of possible overheating and damage to the gel coat (*Fig. 7.4*).

Similar considerations apply when ribs, stiffeners or similar sections are to be added to a moulding. The moulding should be left to harden off before these are added and glassed on. This will ensure that the moulding is rigid enough to resist possible distortion.

*The Inside Surface*

In a majority of cases the "inside" surface is left as stippled or rolled in place. The appearance may however, be improved by the final addition of surface tissue. This can be applied on the "wet" lay-up.

An alternative treatment is to paint the inside surface with a suitable sealer/varnish. This is particularly recommended when that surface of the moulding may be exposed to damp conditions. A sealer/varnish finish cannot be applied until the resin has gelled and hardened.

*Removal from the Mould*

The moulding must be left in the mould for a suitable time to mature properly, so that it can be removed without fear of distortion—see Chapter 1. Even in the case of non-critical mouldings, allow at least 24 hours and preferably twice that time.

It is unlikely, even with a generous application of parting agent, that the moulding will lift easily out of the mould. Usually it will have to be released first by prising away from the sides of the mould. For this job a putty knife or similar tool with a thin flat flexible blade can be useful, but with the edges blunted to avoid scratching either the mould or the moulding. The blade should be worked between the mould and the moulding, and then along, working around the edge. Sometimes it may be necessary to work right round the edge before the moulding becomes free. In other cases, it may free itself almost immediately. There is also the chance that it will stick completely and not come free! In that case, it is a matter of working on freeing it with minimum damage to the mould and moulding— or sacrificing one to save the other, so that a new start can be made.

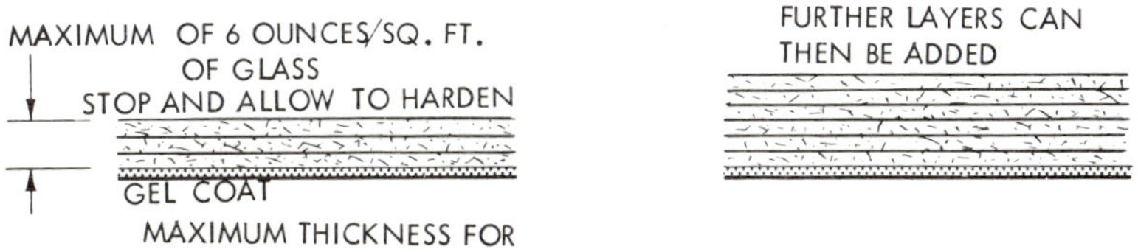

MAXIMUM OF 6 OUNCES/SQ. FT.
OF GLASS
STOP AND ALLOW TO HARDEN

GEL COAT
MAXIMUM THICKNESS FOR
FIG. 7·4 CONTINUOUS WET LAY-UP

FURTHER LAYERS CAN
THEN BE ADDED

# 8. CASTINGS

CASTINGS produced direct from polyester resins are of two main types. The first is where the cast resin is used to embed or "pot" a specimen and forms, in effect, a homogeneous three-dimensional "casing" for the object(s) involved. The other type is where the resin is used as a material for "cold casting", to produce decorative or functional objects—the end result being more or less directly comparable with the forms produced by injection moulding, or other methods of fabricating synthetic resin (plastic) materials.

For the embedding of specimens of insects, small sea creatures, shells, coins, etc.—or for decorative paperweights—clear resins are normally employed, with simple basic shapes. A suitable non-stick mould can be made up from acetate sheet, thick enough to be rigid, with cemented joints (*Fig. 8.1*). Other materials can, of course, be used for making the mould, but may need treatment with release agent.

The resin should be formulated for slow setting—usually based on about 0.5% accelerator and 1 to 2% hardener. This is to ensure minimum exothermic heating during cold-curing, and minimum shrinkage. Ordinary resins can be adjusted for slow setting, but it is preferable to use a special potting resin, which will also ensure a clear, glasslike colour with good optical properties.

The technique is quite simple. Enough resin should be prepared (with catalyst) to fill the mould to a depth of about $\frac{1}{2}$ in. This should be left to gel and harden with the top of the mould covered to exclude dust. Another $\frac{1}{2}$ in. layer of new resin can then be added, and so on. Once suitable height has been built up, the specimen to be potted can be placed on the gelled resin layer, and the mould progressively filled with further layers (*Fig. 8.2*). The main things to avoid are:

(i) Letting dust or foreign matter

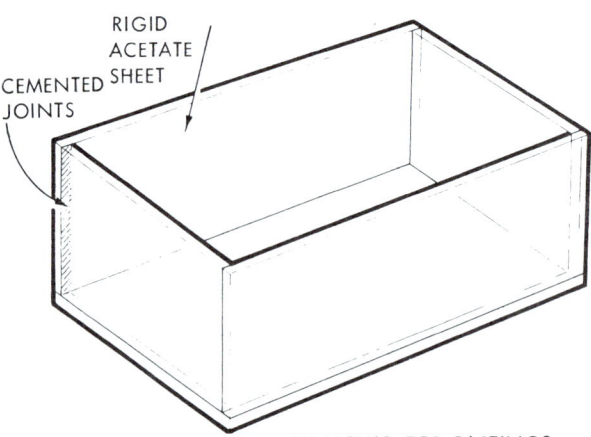

RIGID
ACETATE
SHEET
CEMENTED
JOINTS

FIG. 8·1  HOLLOW BOX MOULD FOR CASTINGS

fall into the mould onto a layer of resin.

(ii) Touching or fingerprinting the soft gel surface.

(iii) Letting air bubbles become trapped in the resin when pouring each layer.

(iv) Pouring too deep a layer at one time, as this can lead to excessive heating (which could damage delicate specimens) and shrinking (which could cause cracks to appear).

(v) Pouring the next layer of resin before the first layer has hardened and cooled down.

When the whole block has hardened completely, it can be removed from the mould. The surfaces can then be cleaned and polished to finish; or even sawn or ground to a different shape and finally polished, preferably with a metal polish or jeweller's rouge and a buffing wheel in an electric drill.

Some slight tackiness may, however, be found on the top surface of the resin. If this shows up as a failing with the resin used it can be cured next time by covering the final surface with cellophane, strapped in place with cellulose tape, to exclude air.

Certain objects need special care when being potted. If the object is flat, like a coin, then there is the possibility of trapping an air bubble underneath it when laying on the gelled layer. This can be avoided by pouring a little liquid resin on to the surface first and pressing and "rocking" the coin down in place in this "puddle" to remove any traces of air before proceeding with filling the next layer.

Specimens or objects which are likely to contain trapped air need a slightly different technique, adjusting the layers around the object so that complete immersion is achieved only after three or four separate layers have been poured. This will allow air to work up through the top part of the specimen and escape.

Objects which are damp, or contain moisture present a special problem. They need drying out or dehydrating completely to achieve entirely satisfactory results. If not, the moisture given out will inhibit setting of the resin. It is generally best to avoid such subjects.

An additional problem arises when the object floats in the resin. Here it is necessary to hold the object down on the bottom layer of resin to avoid it floating out of place when the next layer is poured in. The simplest way of avoiding this is to use a tiny amount of resin to glue the object down to the gelled layer, letting this bond set before pouring on the next layer (*Fig. 8.3*).

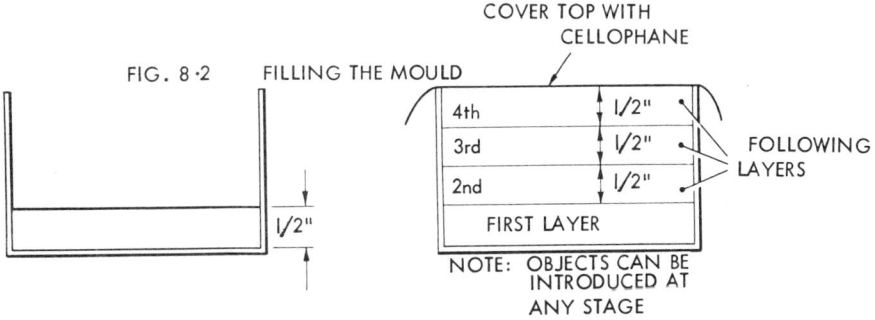

FIG. 8·2    FILLING THE MOULD

COVER TOP WITH CELLOPHANE

| 4th | 1/2" |
| 3rd | 1/2" | FOLLOWING LAYERS |
| 2nd | 1/2" |
| FIRST LAYER | |

1/2"

NOTE: OBJECTS CAN BE INTRODUCED AT ANY STAGE

FIG. 8.3     EMBEDDING A SUBJECT WHICH FLOATS IN RESIN

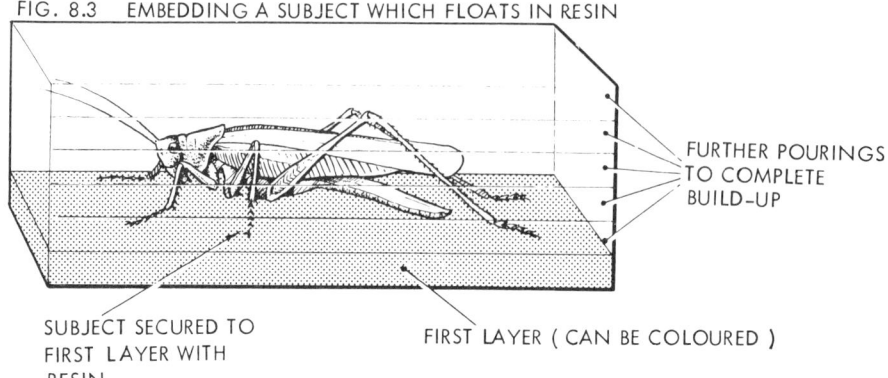

FURTHER POURINGS
TO COMPLETE
BUILD-UP

SUBJECT SECURED TO
FIRST LAYER WITH
RESIN

FIRST LAYER ( CAN BE COLOURED )

### Encapsulation

An exactly similar technique can be applied when potting or encapsulating functional objects, such as an electronic circuit. In this case, though, it is an advantage to mix a filler with the resin to reduce the amount of shrinkage. A preferred filler for electronic encapsulation is mica dust, or similar inert organic filler used in the proportion of between 20 to 30 parts by weight to 100 parts of resin (*Fig. 8.4*).

To obtain the best electrical properties from the resin it is also advisable to post cure the casting by leaving in gentle heat (not more than 80°C) for a period of about ten to twelve hours.

### Decorative and Functional castings

Here the resin used can be clear, coloured with opaque or translucent pigments, or be mixed with fillers. Once again the resin should be of the slow-setting or "casting" type, and

CAST IN 1/2" TO 3/4" LAYERS WITH FILLER RESIN

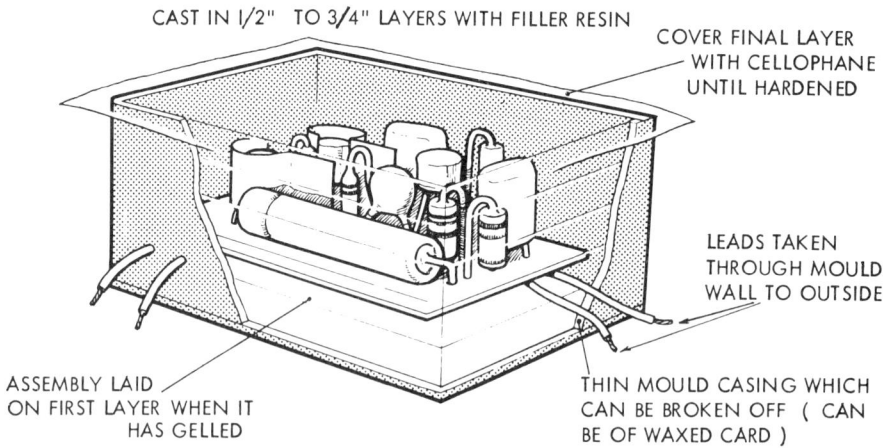

COVER FINAL LAYER
WITH CELLOPHANE
UNTIL HARDENED

LEADS TAKEN
THROUGH MOULD
WALL TO OUTSIDE

ASSEMBLY LAID
ON FIRST LAYER WHEN IT
HAS GELLED

THIN MOULD CASING WHICH
CAN BE BROKEN OFF ( CAN
BE OF WAXED CARD )

FIG. 8·4  ENCAPSULATING AN ELECTRONIC CIRCUIT

in many cases may also be plasticised to reduce internal stresses when setting and eliminate cracking.

Moulds may be of rigid or flexible type. The latter are made from rubber-like materials, readily available, such as cold-curing Silastomer silicone rubber and Vinamold re-meltable rubber. Cold-curing rubbers produce durable, long-lasting moulds with fine reproduction of detail. The material cannot, however, be re-used. Re-meltable rubbers are cheaper, and are particularly suited for making "one off" or limited use moulds, with the advantage that the rubber can then be recovered (by melting down) and used again. The material also sets more quickly, so that a mould can be made ready for use in a minimum of time.

Neither type of flexible mould requires the use of release agents, although as a precaution they can be dusted with French chalk (which should then be smoothed out evenly over the mould surface); or coated with wax emulsion of PVA. Moulds in these materials are prepared by pouring the liquid rubber over a suitable pattern and then allowing to set in the form of a skin. After a suitable thickness has been built up, and the material has hardened, the rubber mould is stripped off the pattern. If necessary, large rubber moulds can be reinforced by bandage strip, scrim, or glass cloth, introduced as a reinforcing layer in building up the skin thickness. The pattern design should avoid undercuts and potential air pockets, and incorporate air vents or risers, if necessary, to prevent air being trapped in the mould when the casting resin is poured in. Instructions for making suitable moulds, and limitations on shapes, are usually supplied with these moulding materials.

Casting is again usually best done in successive layers, the thickness of each layer depending primarily on the bulk of resin involved. It may be necessary to experiment to obtain the best results, and also adjust the proportion of fillers if necessary. It may, for example, be desirable to use a high proportion of fillers to minimise shrinkage and obtain fine detail reproduction, provided a more brittle casting is acceptable. Chapter 4 details the effect of different types of fillers on the mechanical characteristics of the cured resin.

Possibilities with decorative castings are considerable, and by no means restricted to "solid" forms. Other projects include decorative wall panels and wall sculptures, decorative glass panels and tabletops, etc. In other words, casting with polyester resin can be considered not only as a comparable production method to injection moulding, but also to decorative sheet forming and rolling, and vacuum forming, in other types of sheet plastic. Large GRP panels, however, are usually more conveniently produced by moulding—see Chapter 10.

*Dough Moulding*

In this case the casting medium is a resin-filler mix with the addition of chopped up glass fibre strands. The result is a relatively thick paste or "dough" which can be pressed into any suitable mould and left to cold cure, once mixed with catalyst. The proportion of filler should be high enough to ensure freedom from cracking under contraction, although this will also be resisted by the presence of the glass fibres. Once again, however, a slow-setting resin should be used.

Dough mouldings are a quick and simple method of producing solid forms of good strength and durability. With certain shapes it may be advantageous to fit the mould with a closure to which moderate pressure can be applied. This will produce a more compact and uniform product.

*Slush Moulding*

Slush moulding is used to produce hollow coastings in polyester resin. In this case a quick-setting resin is used (mixed with fillers and reinforcement if necessary), poured into a hollow mould which is closed and then rotated rapidly. The resin mix is thrown out onto the mould surface under centrifugal force, where it sets to produce a hollow moulding.

This technique is difficult to reproduce for amateur work as the rotation of the mould really needs to be mechanised to ensure even distribution of the "slush" over the interior surface. Split moulds are also necessary in order to remove the final moulding. For commercial productions of this type the moulds are made of metal and also heated to reduce the setting time.

# 9. BASIC MOULDED SHAPES

## Tray Forms

Only simple moulds are required for producing "tray" shapes, which are usually most effectively constructed from wood on a hardboard or ply base. Where the "tray" is to have a smooth inner surface a male mould is required. Here the basic shape or pattern can be cut from wood block, or a simple built-up box, secured to a base-panel (*Fig. 9.1*). All four bottom corners should be generously radiused.

If the "tray" is to have a plain edge, then the depth of the mould needs to be approximately 1 in. deeper than the final form to allow for trimming off. If the "tray" edge is to be lipped or flanged, then the pattern height is obviously actual size, but the sharp edge should be reduced with the aid of a fillet, as shown in the detail sketches. A further improvement is to build up a separate frame around the basic form, with generous fillets, to produce a rolled-over or beaded edge on the final moulding.

Simple stiffeners are readily incorporated in the mould. Longitudinal stiffeners can be used on medium sizes. These need to be the same depth, and parallel, if the tray is to stand on the base. Larger tray shapes are more effectively braced with diagonal stiffeners. Filletting of the base of stiffeners should not be necessary, provided their depth is kept quite small. Stiffeners added in this way to the mould, of course, produce hollow sections in the base of the final moulding. If raised sections are required—e.g. for photographic dishes—the sections would have to be hollowed out of the base of the mould.

Where the "outside" surface of the tray is required to be smooth, a female mould can be constructed from a simple wood frame laid out on a ply or hardboard base. Fillets are more difficult, although these can be formed in plasticine. No draft or taper is necessary for one-off mouldings, even with deep tray sections, as if the

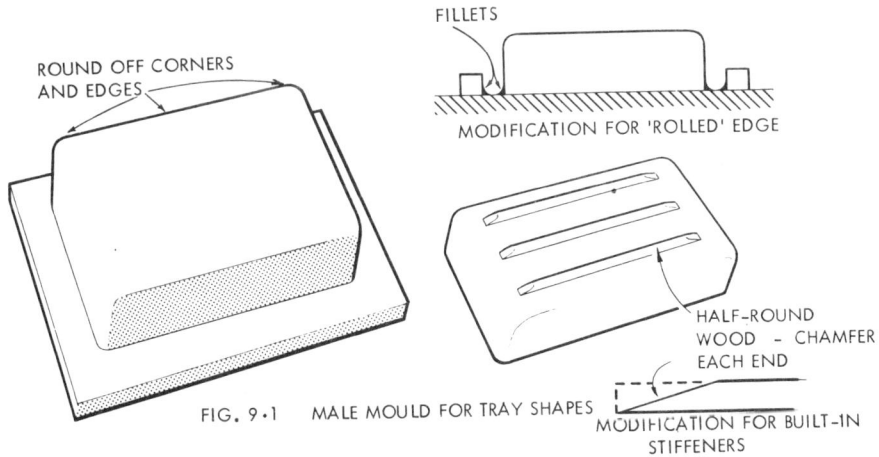

ROUND OFF CORNERS AND EDGES

FILLETS

MODIFICATION FOR 'ROLLED' EDGE

HALF-ROUND WOOD - CHAMFER EACH END

FIG. 9·1    MALE MOULD FOR TRAY SHAPES

MODIFICATION FOR BUILT-IN STIFFENERS

moulding sticks the original framework can be broken away to release.

Lipped edges are easy to accommodate. Stiffening sections can be added directly to the base of the mould. If reverse sections are required, these can be produced by building up the base section with wide strips, as shown in the detail sketches (*Fig. 9.2*). Two layers of $1\frac{1}{2}$ ounce mat should be adequate for tray sizes up to about 24–30 inches (largest dimension) an additional layer can be used on larger mouldings, or where extra rigidity is required.

## Some uses of tray shapes:

Photographic dishes.

Domestic trays (note shapes may be circular or elliptic as well as rectangular).

Complete tabletops (smooth side up, with suitable lip. If stiffening is required, separate stiffeners should be bonded to the underside, or sandwich construction can be used).

Small parts containers and cutlery holders (Divisions can be cut from moulded flat sheet, bonded in place with resin only).

Lids.

Seed boxes (perforations can be drilled in after moulding).

Rigid covers.

Draining boards.

Refrigerator and icebox liners.

Cupboard shelves.

*Box Forms*

Male moulds for producing box mouldings with smooth inner surfaces can be built up from ply or hardboard sides with a thicker wood top. Bracing

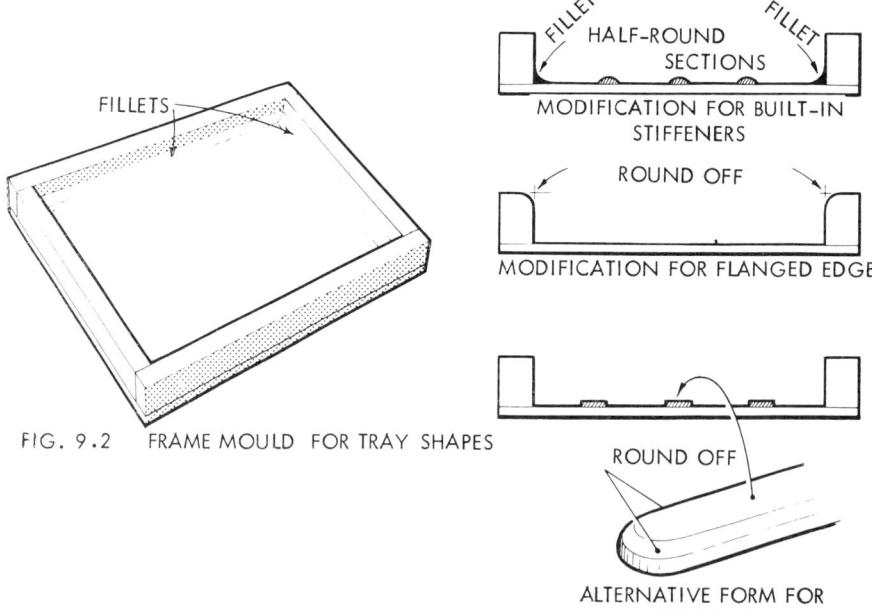

FILLETS

FIG. 9.2    FRAME MOULD FOR TRAY SHAPES

FILLET    HALF-ROUND    FILLET
SECTIONS

MODIFICATION FOR BUILT-IN
STIFFENERS

ROUND OFF

MODIFICATION FOR FLANGED EDGE

ROUND OFF

ALTERNATIVE FORM FOR
STIFFENERS

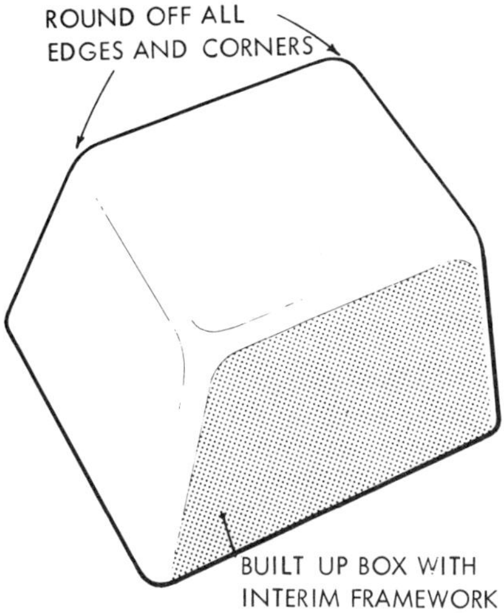

ROUND OFF ALL
EDGES AND CORNERS

5° MINIMUM DRAFT

FIG. 9·3

ALLOW 1"
EXTRA DEPTH
FOR CUT EDGE

BUILT UP BOX WITH
INTERIM FRAMEWORK

FIG. 9·4
MALE MOULD
FOR BASIC BOX
SHAPES

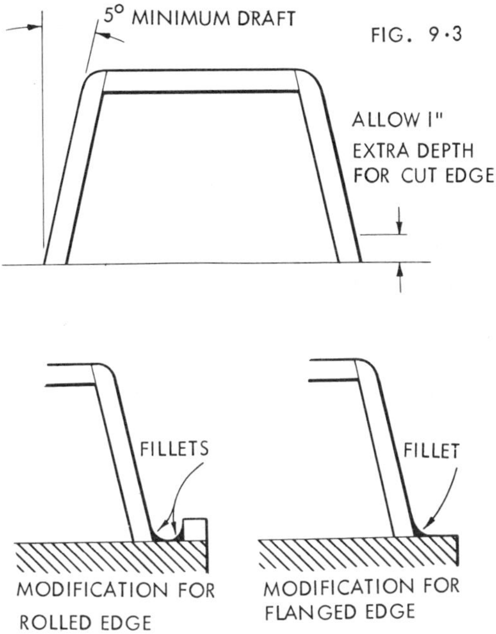

FILLETS

FILLET

MODIFICATION FOR
ROLLED EDGE

MODIFICATION FOR
FLANGED EDGE

pieces should be incorporated in each corner so that all corners, and base edges, can be generously rounded off. At least 5 degrees draft is recommended on the sides to make for easy removal of the moulding from the mould (*Fig. 9.3*). Stiffening sections may or may not be needed on the sides and/or base, depending on the size of the box and its intended function.

Female moulds need construction with external strutting (*Fig. 9.4*). A slightly more generous draft is required, unless only one moulding is required, when the mould can be broken down to release. Filletting of the bottom edge of the mould is a little more difficult because of the limited access, as is smoothing and finishing the mould surface. If a particularly good outer surface finish

is required from the moulding, in fact, it is probably better to make a male mould, and from this a GRP female mould, even for a one-off job.

For making identical, but "handed" boxes, one male mould should be constructed. A GRP moulding is taken off this to form the matching female mould. Further mouldings taken off each of these two will then provide handed pairs of mouldings.

Shapes can, of course, readily be varied when making the original patterns, in which case thick card may take the place of ply or hardboard for curved sides in the construction of a male mould (*Fig. 9.5*). Interior bracing can be used to support the sides to the necessary curvature, as required. Balsa sheet is a convenient material for such strutting, because of the ease with which it can be cut to shape and assembled with quick-drying balsa cement.

Large boxes can easily be built up from moulded flat sheet panels, avoiding the construction of large moulds, although simple jigs may be needed to hold the individual panels in place whilst jointing. Some possible edge joints are shown in *Fig. 9.6*. Choice of edge joint used depends mainly on the size of the construction and its purpose. Mechanical fastening is often an advantage, alternative to, or allied to, bonding with resin and resin glass. In the latter case mechanical fastenings will hold the assembly in place whilst the resin is setting.

**Some uses of moulded boxes (note: lids can be matching "boxes" or "trays", attached with metal hinges):**

Toolboxes.
Containers.
Cases.
Luggage.
Cloches.
Panniers.
Caravan bodies.
Dog kennels (inverted box with cut-out "door").
Battery boxes.
Wheelbarrows.
Tanks.
Ducts (long boxes with open ends).
Sidecars.
Cabinets.
Trailer bodies.
Trucks.

*Shell Mouldings*

Shell mouldings normally have to be made as two separate handed halves, subsequently joined to complete the shell. The exception is a purely cylindrical shell when a single pattern can be used as a male mould—e.g. a length of card tube, as shown in *Fig. 9.7*. To facilitate removal of the tube after the moulding has been laid up on it, the tube should be prepared by slitting carefully down its length to make two or three separate sections, which are then rejoined with masking tape on the inside. After the moulding is completed and has hardened, the

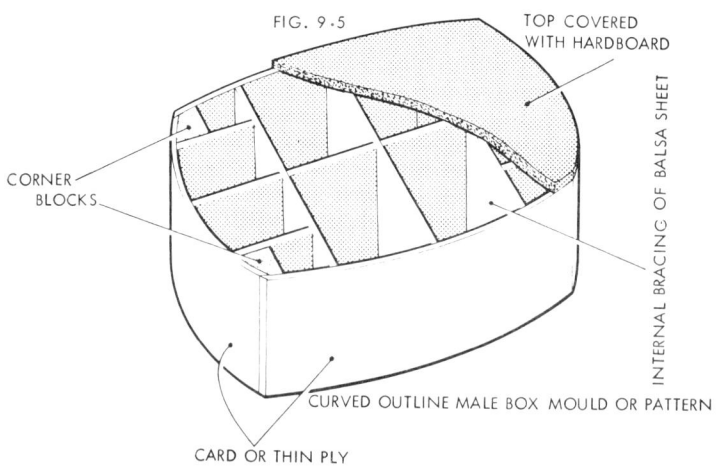

FIG. 9.5

TOP COVERED WITH HARDBOARD

CORNER BLOCKS

INTERNAL BRACING OF BALSA SHEET

CURVED OUTLINE MALE BOX MOULD OR PATTERN

CARD OR THIN PLY

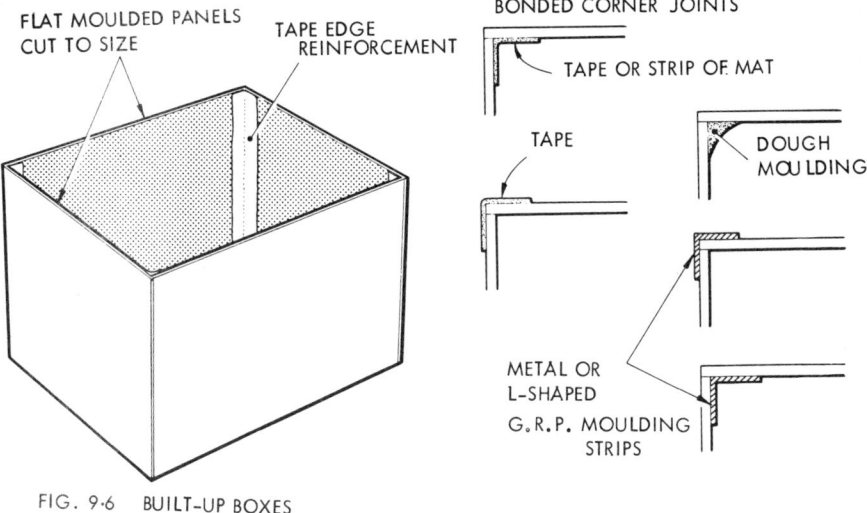

FLAT MOULDED PANELS
CUT TO SIZE

TAPE EDGE
REINFORCEMENT

BONDED CORNER JOINTS

TAPE OR STRIP OF MAT

TAPE

DOUGH
MOULDING

METAL OR
L-SHAPED
G.R.P. MOULDING
STRIPS

FIG. 9·6     BUILT-UP BOXES

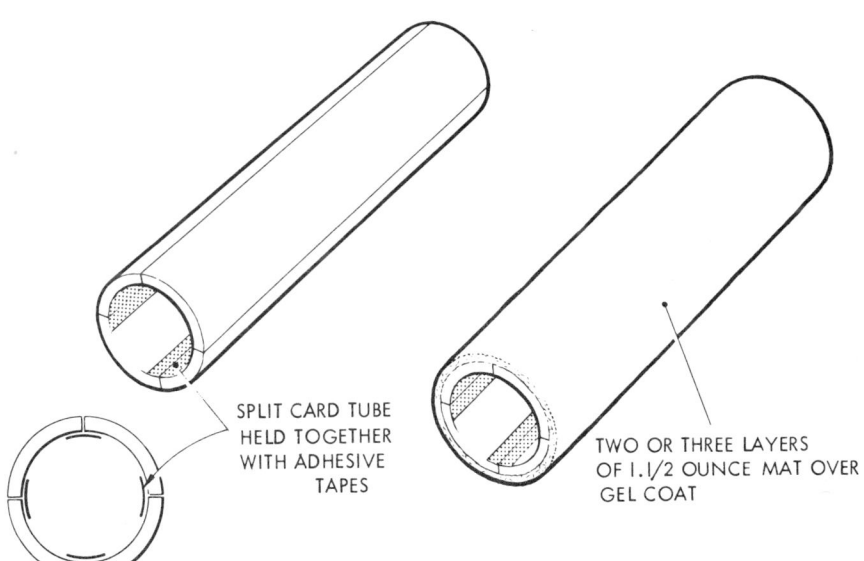

SPLIT CARD TUBE
HELD TOGETHER
WITH ADHESIVE
TAPES

TWO OR THREE LAYERS
OF 1.1/2 OUNCE MAT OVER
GEL COAT

FIG. 9·7     CYLINDRICAL MOULDINGS (SEE ALSO FIG.9·13)

masking tape strips can be pulled off through the inside of the tube, when it should be possible to spring the card tube sections apart and withdraw them.

A similar technique can be applied to straight tapered shell shapes. In this case a male mould is constructed around a tapered core piece, as shown in *Fig. 9.8*, held in place with a locking plate at each end. After the lay-up is complete the locking plates are removed and the core piece withdrawn. The outer pieces of the mould can then be sprung free and withdrawn.

Since both of these methods utilise a male mould, the shell mouldings are of smooth bore, but with a rough outer surface. Shell mouldings with a smooth outer surface require the use of handed female moulds, each producing a half shell. In this case a plug is constructed first, then separated down the middle, and a female GRP mould laid up on each.

A plug for splitting should be made up from two separate, but identical,

blocks or sub-assemblies which are only lightly joined together. This will enable them to be worked as one, for final shaping, but easily separated along the true centreline (*Fig. 9.9*). Further additions must then be made to these two half shell patterns, according to the method to be used for joining the half shell mouldings.

If a butt joint is to be used, then each pattern must be extended to provide extra "width" of moulding, so that it can be trimmed back to a clean edge (*Fig. 9.10*). A bonded joint of this type needs very accurate trimming of the cut edge of each moulding, if a perfect joint is to be obtained. The joint can be strengthened with glass tape applied on the inside of the assembly, as shown, provided this joint line is accessible. For straight butt joints it may be desirable to increase the thickness of the moulding locally in the region of the edge, which can be done quite simply when laying up in the female mould. No modification of the pattern is required for this.

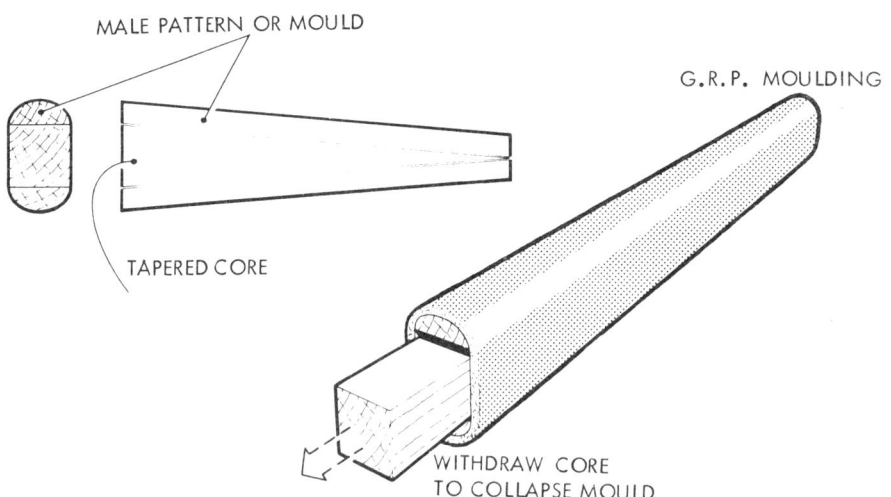

MALE PATTERN OR MOULD

G.R.P. MOULDING

TAPERED CORE

WITHDRAW CORE
TO COLLAPSE MOULD

FIG. 9·8     STRAIGHT TAPERED SHELL MOULDINGS

Gaps in a butt joint can also be filled with resin/filler. If the moulding is coloured, then colour matching is important when gap filling. The same pigment paste used in resin/filler will need careful matching for shade. Pigmented (but not activated) resin saved from the original gel coating will not give a match when mixed with filler powder because of the whitening effect of the latter, unless of course the colour is white to start with.

Flanged joints are much stronger and easier to fit up, but not always acceptable from the point of view of appearance. Some modification of the pattern is needed to allow for local thickening, and the formation of the flange—see *Fig. 9.11.*

Turned-in flanges are not a practical proposition for moulding with the shell although they can be produced if the shell is flexible enough to be sprung out of the mould (*Fig. 9.12*). The main difficulty then is getting a "sharp" edge in the inside corner, which is the least accessible when laying up, and the most visible when the shells are joined. If turned-in flanges are to be used, they are best added as separate sections after the moulded shells have been removed from the female mould.

Various other methods of joining half shells neatly and effectively can often be devised, depending on the size and shape involved. Such problems present a challenge to individual ingenuity.

**Some uses for Shell mouldings:**

Pipes and ducts.
Model aircraft fuselages.
Streamlined or rounded form tanks.
Coal hods (tapered cylindrical shell
  with separately fitted end moulding).

**Some uses for Half Shells** (made

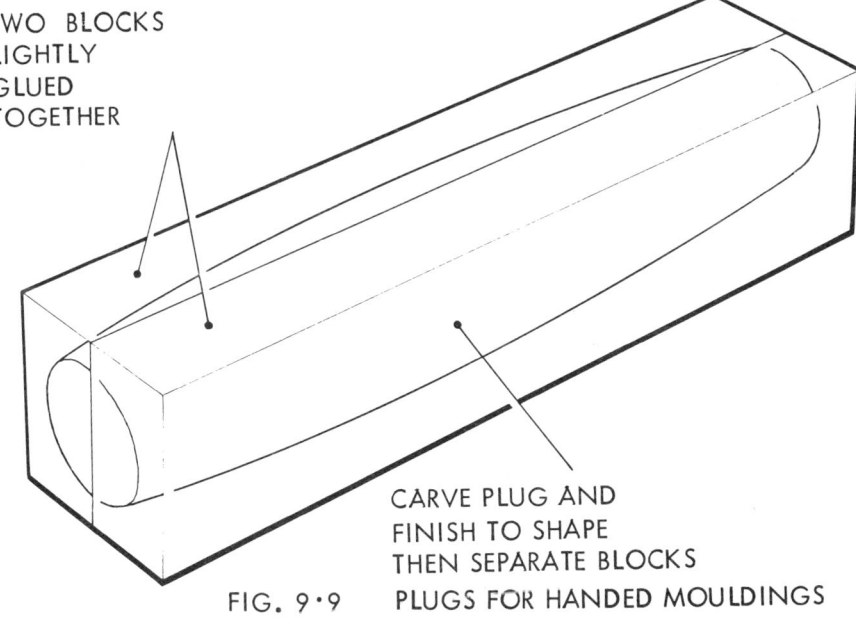

TWO BLOCKS
LIGHTLY
GLUED
TOGETHER

CARVE PLUG AND
FINISH TO SHAPE
THEN SEPARATE BLOCKS

FIG. 9·9    PLUGS FOR HANDED MOULDINGS

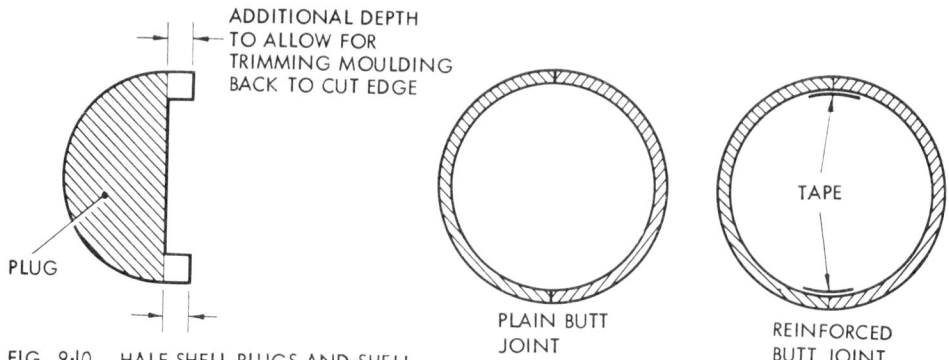

FIG. 9·10   HALF SHELL PLUGS AND SHELL

directly off a male pattern; or from a female mould taken off a male pattern, depending on which is to be the smooth side):

Scooter legshields.
Machine guards.
Cloches.
Fairings.
Reflectors.
Shades.

Polyester resin with glass fibre reinforcement is particularly success-ful for making pipes and ducting of various kinds. The ease with which it can be formed into complex pipe shapes with difficult junctions, and the fact that it is resistant to most acids and alkalis, in addition to being proof against more ordinary substances such as oil and water, makes it a good choice for such work.

Straight lengths of pipe can be made round split mandrels by wrapping with strips of cloth or tape impregnated with resin. A simple

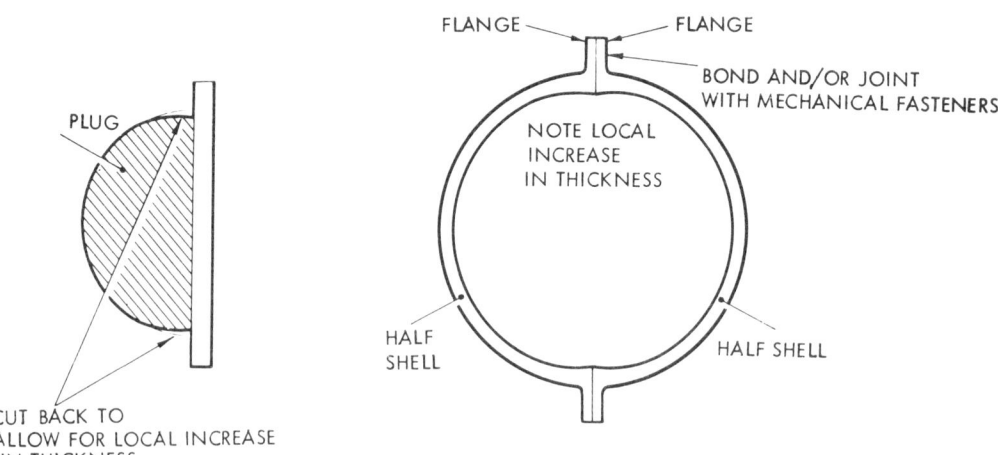

FIG. 9·11   FLANGED HALF SHELLS

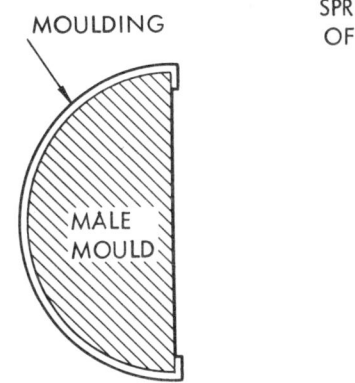
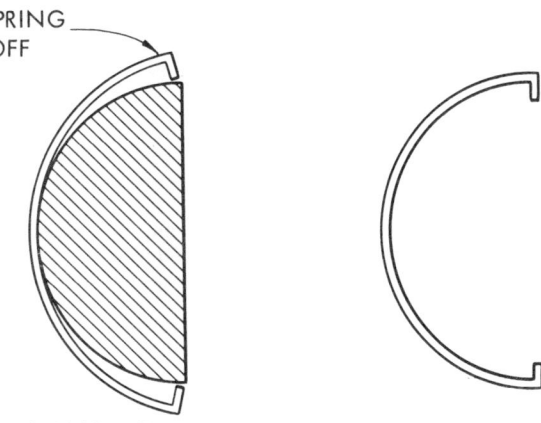

MOULDING

MALE MOULD

SPRING OFF

FIG. 9·12    HALF SHELLS WITH TURNED-IN FLANGE
CAN ONLY BE PRODUCED ON A MALE MOULD

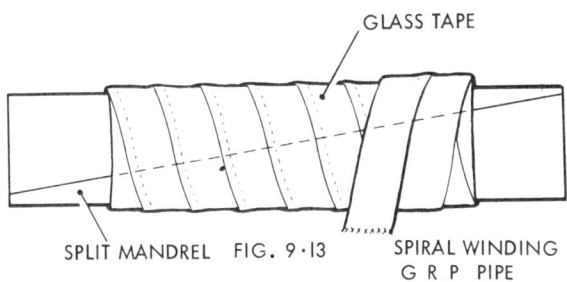

GLASS TAPE

SPLIT MANDREL    FIG. 9·13    SPIRAL WINDING
G R P PIPE

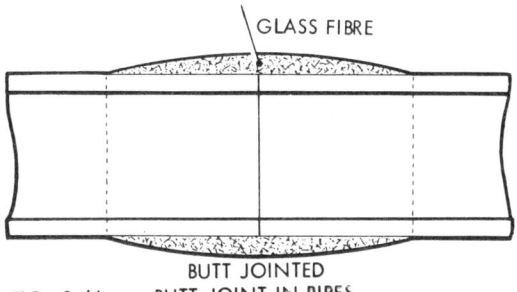

GLASS FIBRE

BUTT JOINTED
FIG. 9·14    BUTT JOINT IN PIPES

example of this is shown in *Fig. 9.13*.

Larger diameter pipes can also be made inside a length of metal tube by placing glass mat, cut to overlap slightly, inside the tube; then pouring in a quantity of resin and rotating the tube rapidly, so that the centrifugal force causes the resin to impregnate the mat. Heat applied to the outside of the tube assists the curing process, and subsequent release is relatively easy, as the moulding shrinks away from the metal tube.

Pipes with simple bends in them can be made on the outside of flexible moulds such as plastic tubing and "Vinamold" mouldings. Although the resin impregnated tape or cloth wrapped round the former can be kept in close contact, and given a tidier surface, if it is wrapped with strips of cellophane. Failing this, it is essential not to allow too much free resin on the surface of the cloth, as it tends to gather in blobs and give a poor appearance.

Ducting of a complex shape can be formed over a mould made of a material with a low melting point, and after the laminate has cured, it can be heated in an oven and the mould melted out. Paraffin wax and some of the special fusible metal alloys are suitable for this kind of job.

Pipes or ducts can be joined together simply and effectively by butt joining them and binding with several turns of glass tape impregnated with resin (*Fig. 9.14*). The ends to be joined should be first cleaned thoroughly by scraping or by sanding. When the resin sets it contracts sufficiently to make a very firm joint.

Emergency repairs to burst water pipes can be successfully made in the manner described above, although it is as well to remember that burst pipes usually occur in cold weather in cold places and it is best to apply some local heat to the joint to ensure that the resin cures properly.

# 10. ROOFLIGHTS AND TRANSLUCENT PANELS

FOR producing "transparent" or translucent moulded panels a clear light-stabilised resin should be used (and amine type accelerators avoided), together with "E" grade glass for the reinforcement. The description "transparent" is applicable in the case of plain optical resins since a light transmission of the order of 70 to 80% can be achieved. Coloured translucent panels can equally well be produced, using suitable pigments or pigment-dyes in the resin. Moulding technique is extremely simple as no special moulds are required.

*Flat Sheets*

Any suitably smooth and flat surface will do as a mould, such as a sheet of new hardboard (used shiny side up). This should be several inches larger all round than the size of moulded panel required.

The hardboard panel should be covered with a single sheet of cellophane, stretched in place, taut and free from wrinkles, taken around the edges and taped in place to the back with adhesive tape.

Cut glass mat to the required panel size. For small panels, a single piece of $1\frac{1}{2}$ ounce mat will be adequate. For medium size panels, a single piece of 2 ounce mat can be used. For large panels, two pieces of $1\frac{1}{2}$ ounce mat should be cut.

Mix up the resin, with or without pigment and paint a generous layer on to the cellophane covered hardboard surface. Lay the mat in place on top of the resin, making sure that it lies flat. Stipple in place lightly with a brush and then cover with a second sheet of cellophane (*Fig. 10.1*). Use a roller to consolidate the resin/glass and work out all traces of air bubbles. To produce a uniform and even smoother top surface a second piece of hardboard can be laid in place over the cellophane and the "sandwich" further run over evenly with a roller. Remove this piece of hardboard and check that the panel appears sound throughout, viewing through the cellophane layer. Leave the cellophane in place until the resin has set and hardened. The top layer of cellophane can then be peeled off, and the moulded panel lifted off the cellophane covered hardboard.

*Corrugated Panels*

The moulding procedure is similar, except that corrugated sheeting (e.g. asbestos or iron) is used in place of hardboard. The width of glass mat needed to produce a required width of corrugated sheet can be found quite simply by running a piece of adhesive tape across the sheet, sticking to the corrugations (*Fig. 10.2*). Strip off, straighten out and measure the corresponding "flat" width.

The corrugated sheet pattern is again covered with a single sheet of cellophane, draped in position. This should be generously oversize in width to allow it to be drawn down into the corrugations and still leave some spare area each side. Paint resin directly on to the cellophane and follow up with the glass mat. This is then pressed and rolled down in place. Again a covering with a second sheet of cellophane, followed by placing an identical piece of corrugated sheet over the top and pressing down can be used to form the final shape, once all air bubbles have been removed by stippling.

If difficulty is experienced in laying the glass fibre in place and forming it into the corrugations, an alternative technique can be tried. Lay the first

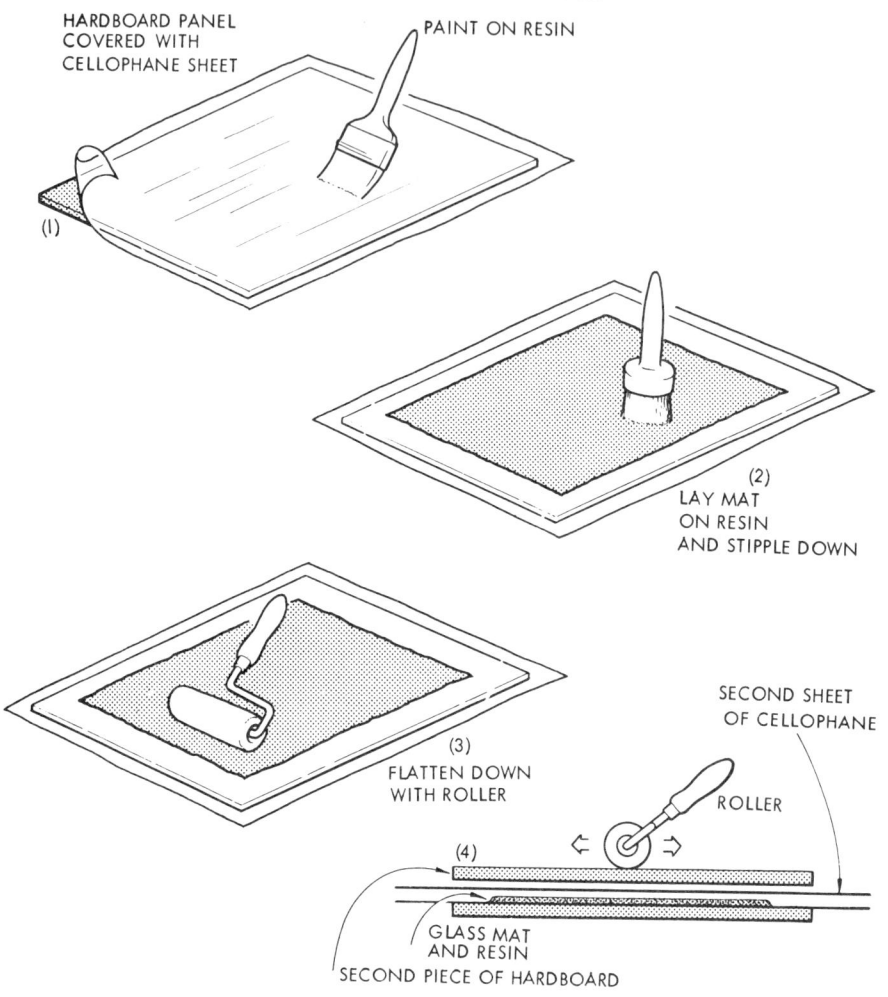

FIG. 10·1    MOULDING FLAT SHEETS

FIG. 10·2    MEASURE THE 'FLAT' WIDTH OF CORRUGATED SHEET

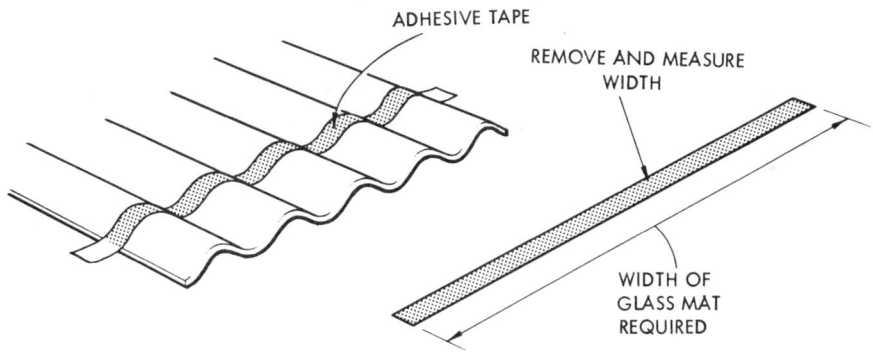

ADHESIVE TAPE

REMOVE AND MEASURE
WIDTH

WIDTH OF
GLASS MAT
REQUIRED

FIG. 10·3    MAKING CORRUGATED SHEET MOULDINGS

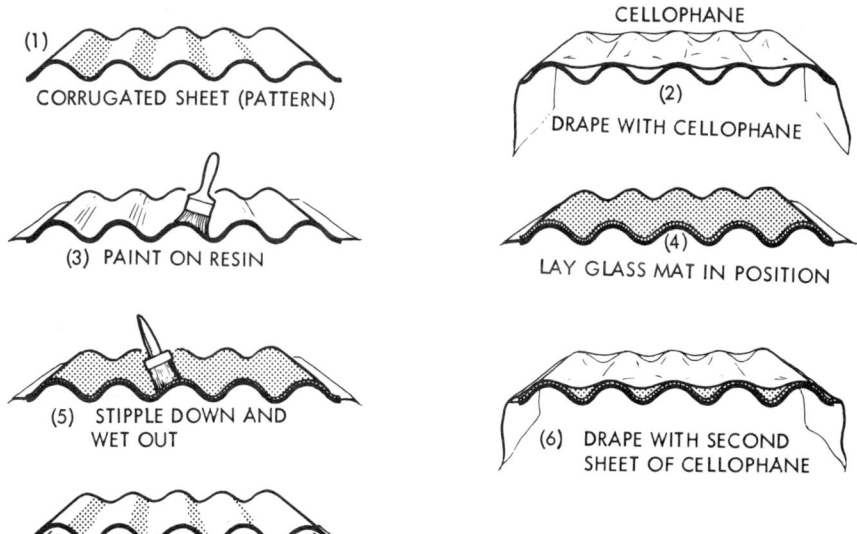

(1) CORRUGATED SHEET (PATTERN)

CELLOPHANE

(2) DRAPE WITH CELLOPHANE

(3) PAINT ON RESIN

(4) LAY GLASS MAT IN POSITION

(5) STIPPLE DOWN AND WET OUT

(6) DRAPE WITH SECOND SHEET OF CELLOPHANE

(7) LAY SECOND PIECE OF CORRUGATED SHEET IN PLACE AND PRESS DOWN UNIFORMLY

cellophane sheet out on a flat surface (e.g. a piece of hardboard), first making sure that the cellophane is wide enough for the job. Paint resin on the flat cellophane and lay the glass mat on top, stippling down until thoroughly wetted. Then transfer cellophane and resin to the corrugated sheet for final forming.

Panel edges should not be trimmed to shape until the sheet has fully hardened. The panel can then be cut to final size with a hacksaw or metal saw.

### Some uses for Flat Sheets:

Small rooflight panes.
"Unbreakable" windows for garden sheds, etc.

Obscure window panes.
"Stained glass' window panes (translucent colours).
Heat resistant mats.
Heat-resistant tabletops.
Decorative "see-through" table tops.
Built-up box constructions.
Shelves (with added stiffeners).
Lightweight doors (with added stiffeners).
Garage doors (from separate panels, joined).

### Some uses for Corrugated sheets:

Rooflights.
Roofing for garden rooms, sheds, outhouses, garages, etc.
Patio windbreaks.

# 11. GRP TANKS

TANKS for holding liquids are really a special form of "box", although the actual shape may be dictated by styling requirements, or the shape in which they are to fit. Logically the smooth side of the GRP should come on the inside of the tank, which will involve the use of two male moulds and two separate mouldings. The joint line should be positioned to come above the contents level, if possible when the two separate mouldings can comprise the basic tank section itself, and a shallow lid—e.g. see *Fig. 11.1*.

A flanged joint is to be preferred if the tank is to be sealed as this will give better control of the bonded joint, which can be clamped up during setting, or further secured with mechanical fasteners. Aesthetically, how-ever, a lap joint may be preferred, particularly if the tank is of stylised shape. Particular care must be taken over the bonding together of the two mouldings if the joint line has to come below the normal level of the tank contents—e.g. see *Fig. 11.2*.

The possibility of using built-up construction, employing flat moulded sheets, should not be overlooked where a square, rectangular or wedge-shaped tank is required. Joints can be reinforced inside and out with glass tape or mat or cloth strips—see *Fig. 11.3*. Whilst this eliminates the need for making special moulds, it does place a premium on accurate cutting to shape of the individual panels.

Fillers, mounting lugs, vent pipe

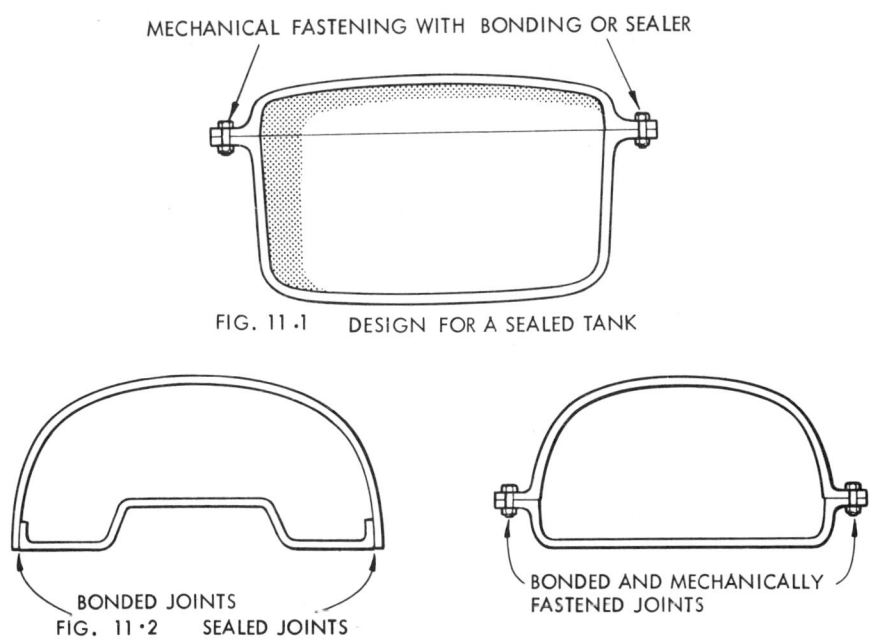

MECHANICAL FASTENING WITH BONDING OR SEALER

FIG. 11.1    DESIGN FOR A SEALED TANK

BONDED JOINTS
FIG. 11.2    SEALED JOINTS

BONDED AND MECHANICALLY
FASTENED JOINTS

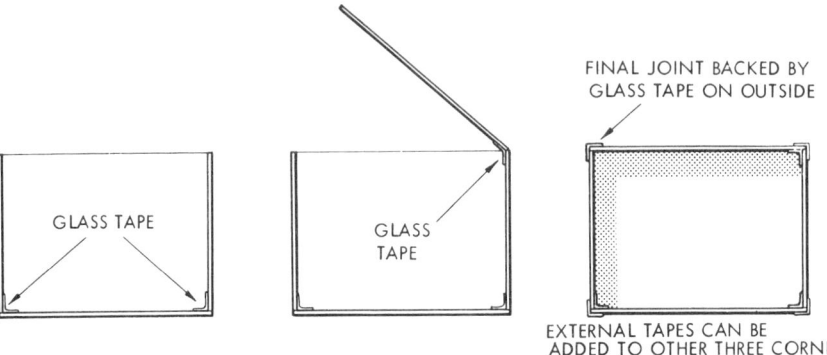

FINAL JOINT BACKED BY
GLASS TAPE ON OUTSIDE

GLASS TAPE

GLASS
TAPE

EXTERNAL TAPES CAN BE
ADDED TO OTHER THREE CORNERS

FIG. 11·3    RECTANGULAR TANKS FROM MOULDED SHEET PANELS

fittings, brackets, etc., can readily be bonded in place with an overlay of glass mat.

In the case of tanks built to hold water or chemical solutions, the finished tank should be left to age for at least two to three weeks before being put into use in order to reach maximum resistance to water absorption and chemical attack. With petrol tanks an even longer ageing period is recommended, if possible. With petrol tanks, too, particular care should be taken during the lay-up to ensure a perfect gel coat of good thickness. Perfect sealing of the surface is important, as if stray glass fibres can become loosened they can be carried by the fuel into the carburettor and cause blockage. A slightly flexible resin for the gel coat might be an advantage to eliminate any possibility of the gel coat crazing or cracking under vibrational loads it might experience in service.

### Tank Repairs

Glass fibre is a very good medium for repairing leaky tanks of all kinds. The main requirement is that the tank is empty and quite dry. Loose corrosion and rust should be cleaned off the affected area with a wire brush. The leak can then be repaired with a "patch" of glass fibre mat, applied with resin.

In the case of a tank which can be opened the "patch" repair is best applied on the inside, extending over as wide an area as necessary (*Fig. 11.4*). If perforation of the tank is very prominent, the outside surface should be backed up with a piece of cellophane covered card or hardboard, held in place with adhesive tape. The patch thickness can be quite generous —say three layers of $1\frac{1}{2}$ ounce mat for an average repair. The resin mix may have to be adjusted to suit the air temperature, if the tank is repaired *in situ*. Hardening off can be speeded by placing a suitable heater near the tank when the repair is completed. In the case of a repair to a cold water tank it is obvious that the tank is needed back in service as soon as possible. Use heat to post-cure and allow a minimum of twelve hours hardening time, if possible.

Sealed tanks, such as fuel tanks, must be patched on the outside, unless they have an inspection panel which can be removed and gives access to the damage from the inside.

The procedure is essentially the same, but the tank must be fully emptied and the patch area *completely degreased*, otherwise the resin will not adhere. This can be done by scrubbing with trichlorethylene, carbon tetrachloride, or a strong detergent solution. In the latter case, check that the damaged area is dried out thoroughly before proceeding with the repair. Even when dry on the outside, some moisture could remain inside the tank, where it has entered through the original damage.

FIG. 11·4

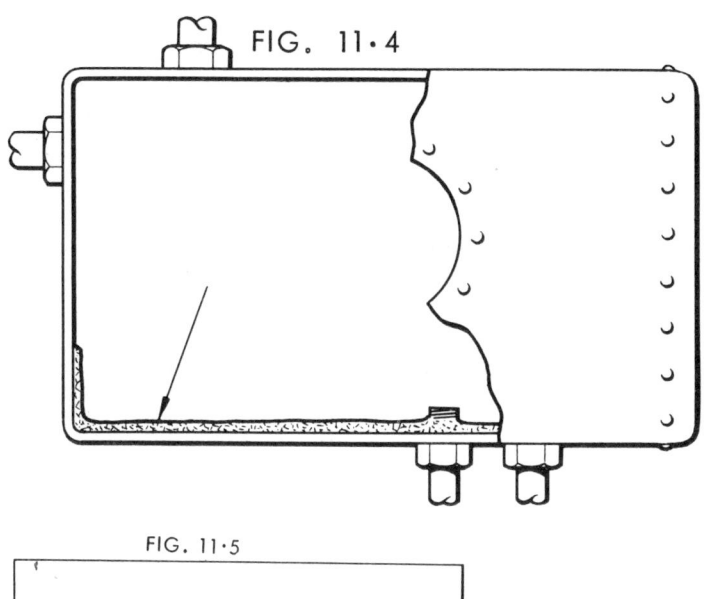

FIG. 11·5

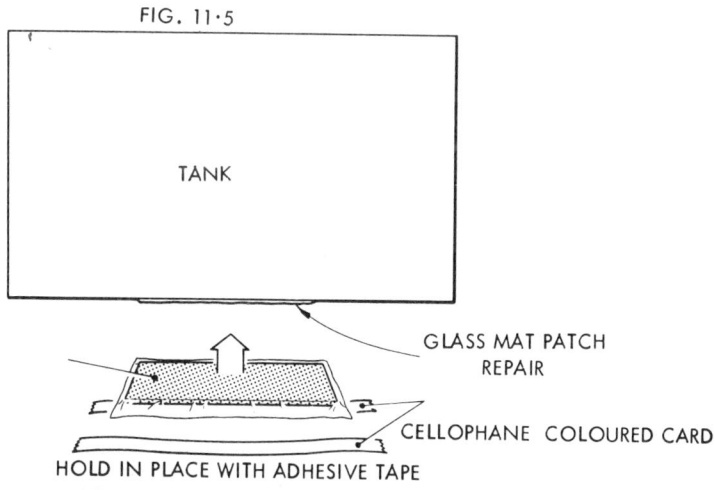

TANK

GLASS MAT PATCH
REPAIR

CELLOPHANE  COLOURED CARD

HOLD IN PLACE WITH ADHESIVE TAPE

IN SITU REPAIR TO BOTTOM OF A TANK

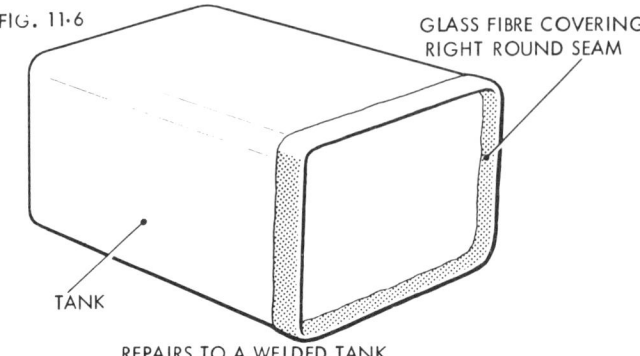

FIG. 11·6

GLASS FIBRE COVERING
RIGHT ROUND SEAM

TANK

REPAIRS TO A WELDED TANK

If the repair has to be made *in situ* and is on the bottom of the tank (as is usually the case with a car petrol tank), the patch may have to be supported in place with a piece of cellophane covered card or hardboard, strapped to the bottom of the tank with adhesive tape (*Fig. 11.5*). The resin should have enough tackiness for the patch to be positioned properly, working from underneath. The additional back-up is to ensure that it does not sag or pull off under its own weight whilst setting.

Large metal storage tanks of welded construction may eventually fail at the seams. A replacement tank can be very expensive and so a repair job is usually worthwhile, even if all the seams of the tank are suspect as a consequence, or because of age. In this case the repair or overlay or glass fibre should be continued right round the complete length of seam, not confined to the damaged area (*Fig. 11.6*). In extreme cases it might even be considered advisable to clad the whole of the tank with glass fibre. In that case the original tank merely becomes an imperfect "liner", but any further deterioration will not lead to a new leak developing. The liner will, however, be in direct contact with the contents and corrosion products may be withdrawn with them. The chances of this happening can be minimised by flushing out the tank thoroughly with an inhibiting fluid.

# 12. GRP BOATS

BOAT construction is probably the biggest single application of GRP. Unfortunately, it is also the least suited to amateur construction, primarily because hull moulding virtually demands the use of a female mould. This means the construction of a full size plug first, on which the mould is laid up (invariably in GRP). The method is both lengthy, and expensive, for one-off productions.

It is, however, possible to construct a female mould direct, provided the builder is reasonably skilled as a carpenter and can interpret hull line drawings. Thus the job should be within the scope of any modeller wanting to attempt the construction

of a full size dinghy or runabout in GRP.

The simplest hull form to attempt is the deep-vee powerboat hull form with constant deadrise; this being originally designed to be constructed from flat ply panels, and thus free from compound curves. It is possible to reproduce such a hull form from five flat panels of hardboard, suitably assembled and braced in an external framework, as shown in *Fig. 12.1*.

Hull lines can be taken off any suitable scale drawing of a full size hull. The external frames should be erected first, made from any suitable wood, screwed rigidly together. The skin panels are then attached to these

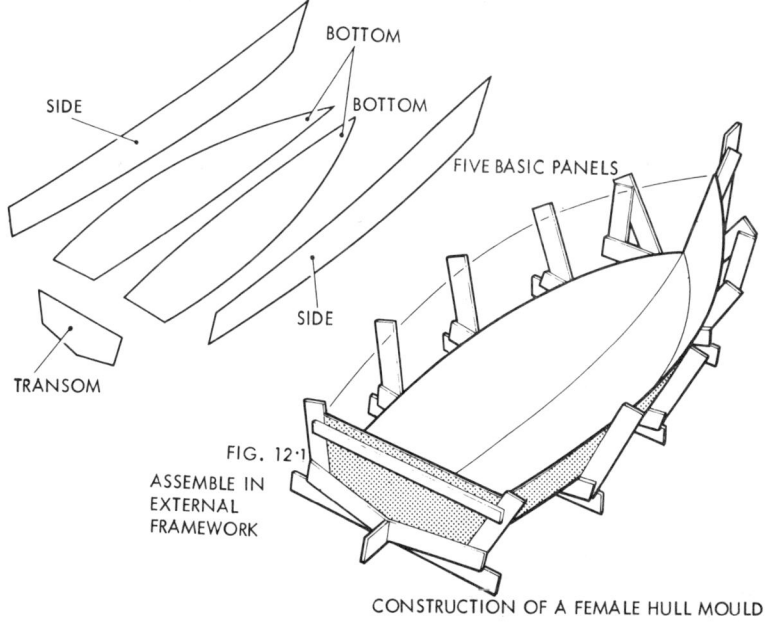

FIG. 12·1

ASSEMBLE IN
EXTERNAL
FRAMEWORK

CONSTRUCTION OF A FEMALE HULL MOULD

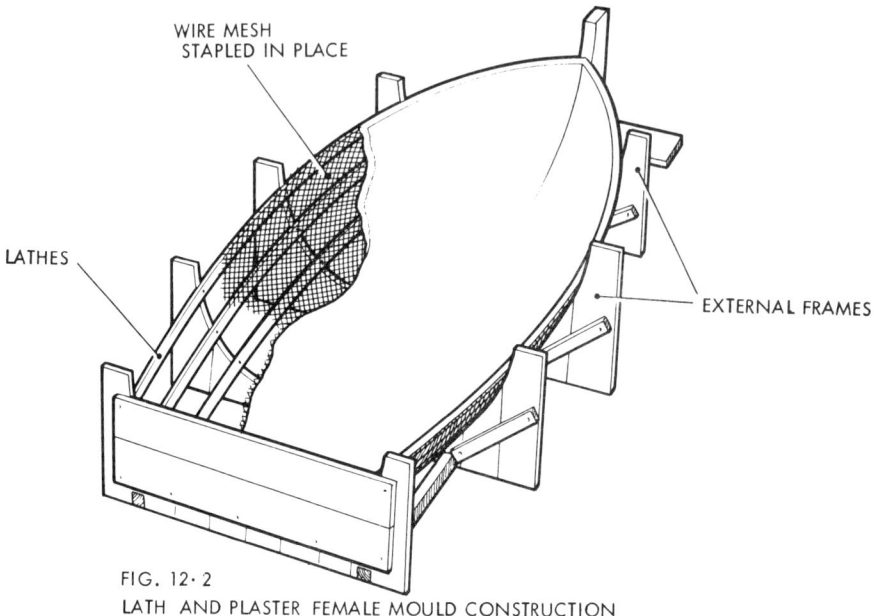

WIRE MESH
STAPLED IN PLACE

LATHES

EXTERNAL FRAMES

FIG. 12·2
LATH AND PLASTER FEMALE MOULD CONSTRUCTION

frames, using a minimum number of countersunk head screws, with the head pulled down below the surface of the hardboard. Bottom panels can be finally fitted up by cut-and-try methods. Gaps or bad fits are not necessarily important as these can be filled in as necessary. The main object is to complete a "hollow" hull shape with the surface of the mould completely uncluttered so that it can be made good as necessary, smoothed right off and polished in the manner of a female mould. Small deviations in shape from the original drawings could be quite acceptable, provided the hull form produced remained perfectly symmetrical.

No attempt should be made to "cut in" chines and longitudinal strakes on the bottom (essential for stability of a dee-vee hull). This would only make

the mould construction unnecessarily tedious, as well as making lay-up in the finished mould more difficult. Such items can be mounted later on the finished moulding.

Various refinements are possible on this relatively straightforward method of female mould construction. Thus the deadrise of vee-angle could be greatly reduced, and the sharp angle between bottom panels and sides could be rounded off with a fillet of plaster, producing more of a round bilge form, suitable for a sailing dinghy or utility runabout rather than a speedboat.

True round bilge forms are more difficult to mock up since this would normally call for several external, and accurately plotted, frames on which are fastened laths, bent to the right curvature. These laths would need

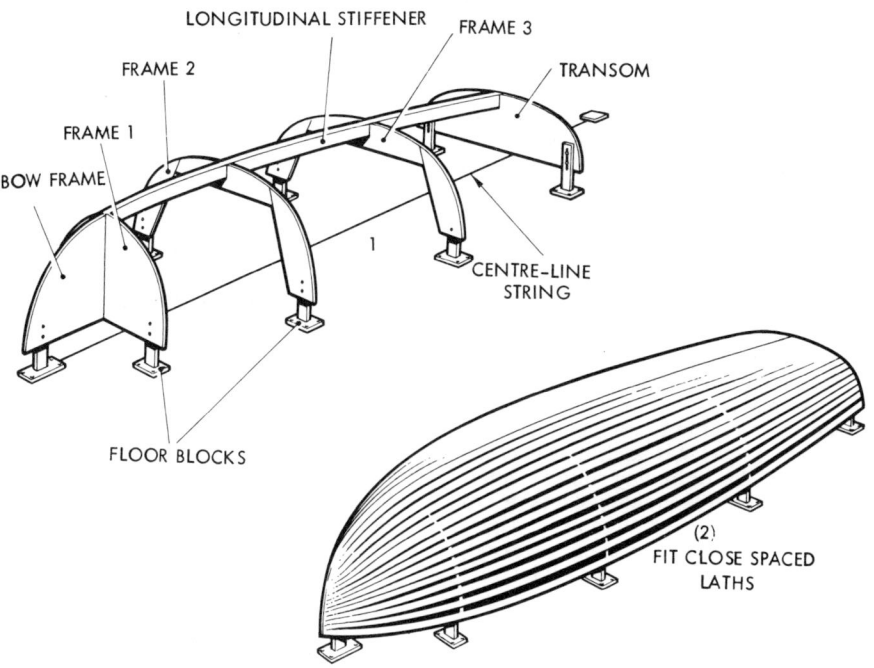

LONGITUDINAL STIFFENER

FRAME 3

FRAME 2

TRANSOM

FRAME 1

BOW FRAME

1

CENTRE-LINE
STRING

FLOOR BLOCKS

(2)
FIT CLOSE SPACED
LATHS

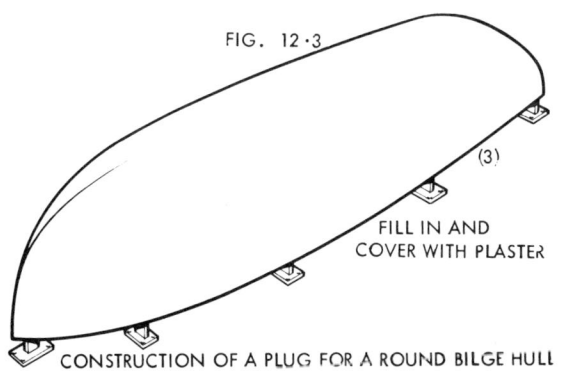

FIG. 12·3

(3)

FILL IN AND
COVER WITH PLASTER

CONSTRUCTION OF A PLUG FOR A ROUND BILGE HULL

close spacing, especially in regions of double curvature, and tapering off at the bows and stern where they would tend to become crowded together (*Fig. 122*).

Such a mock-up could be completed, as a mould, by covering with wire mesh, held in place with small staples, followed by draping in hessian soaked in plaster. Once this had hardened in position, a rough coating of plaster trowelled on and smoothed out, followed by a smooth final plaster coating completes the mould to the stage where it is ready for final smoothing, sanding down and sealing.

Moulds of this type may take less time to make than a plug, from which a female mould is to be taken in GRP. A plug, however, has the advantage that it is "the right way round" rather than a three-dimensional "image" of the actual hull, and thus it is easier to appreciate the shapes involved, where they may need modification, and so on. The other two advantages of a plug are:

(i) Compound curves can more readily be incorporated, particularly to improve bow shapes.

(ii) Appendages such as chines, keel, external strakes, etc., can readily be added to the plug, and thus incorporated in the GRP mould when made.

*Fig. 12.3* shows a suitable method of making a plug for a round bilge hull. A plug for a hard chine hull could be made directly from flat panels assembled on a simple internal framework.

Construction of a plug for a female mould can be eliminated if a suitable subject can be found ready-made. For example, an existing hull of suitable size and shape can be used directly as a plug. The only work involved then is stripping off appendages and fittings which would interfere with the production of a suitable GRP mould, when laid up on it; and making good, sealing and finishing the surface of the hull to a suitable standard for taking a moulding off. This can be the major work involved in such a project. Starting with an old hull, though, which will not be used again, various modifications can be built on to the original hull to improve the shape, again using the simplest and cheapest materials available for the job.

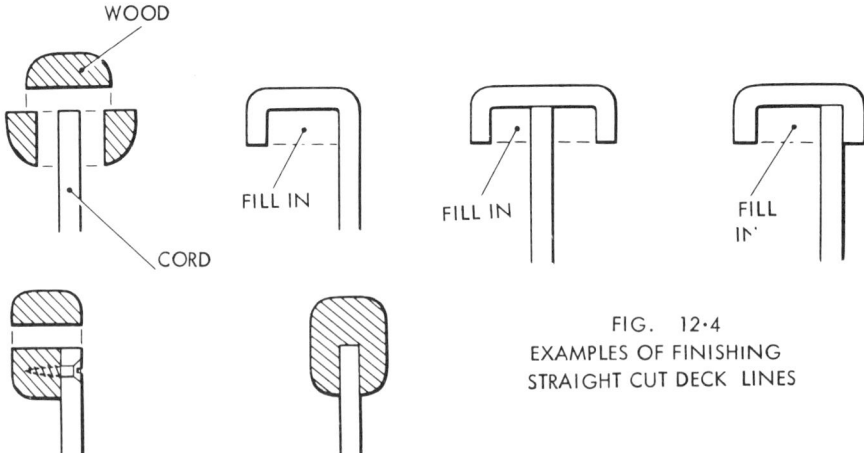

FIG. 12·4
EXAMPLES OF FINISHING
STRAIGHT CUT DECK LINES

A further possibility worth investigating is the *hire* of female moulds for GRP hull construction. These are available in various sizes and designs, the cost of hire ranging from about £25 per week upwards.

Moulds for hull designs must allow for trimming of the final moulding, and how the top line of the hull is to be finished. *Fig. 12.4* shows various methods of finishing off a plain trimmed hull moulding, such as might be used in small open boats. The incorporation of a "rolled over" edge in such cases is good practice, adding considerable stiffness and rigidity to the top lines of the hull. Such "rolled-

over" sections should always be filled.

Where the hull is to be fitted with a deck, then a flanged edge is preferable so that the hull and deck moulding can be screwed, stapled, bolted or riveted together (*Fig. 12.5*). Mechanical fastening is preferred to bonding, although the two may be combined (or sealer used instead of adhesive) to ensure a watertight joint. There are alternative methods, such as the addition of separate gunwale and inwale pieces to provide a seating for a deck, and a number of possible variations are shown in *Fig. 12.6*. This by no means exhausts the possibilities.

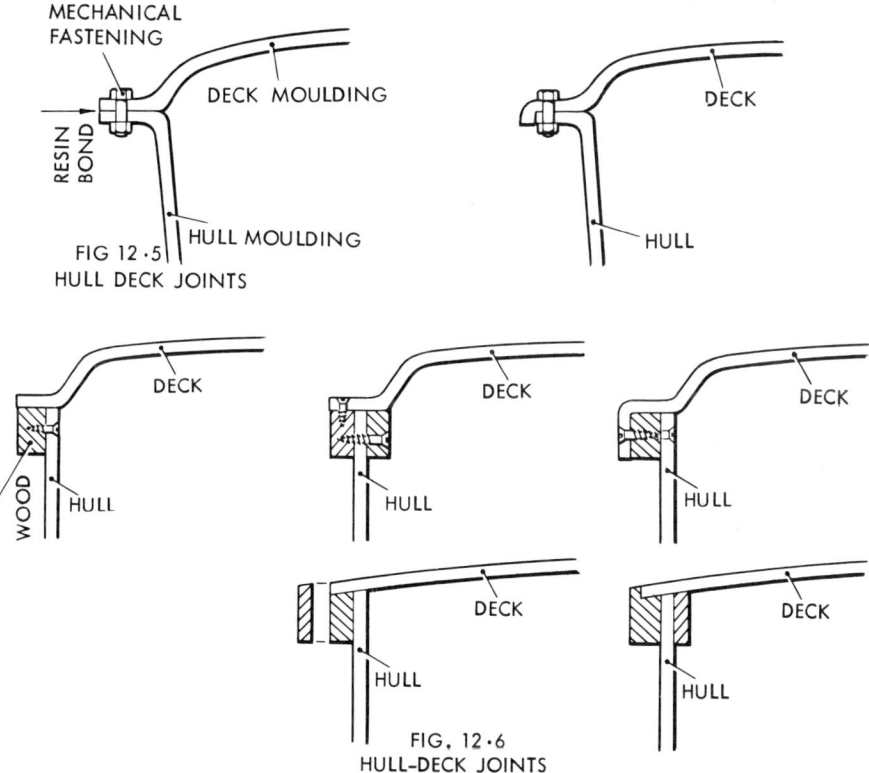

FIG 12·5
HULL DECK JOINTS

FIG. 12·6
HULL-DECK JOINTS

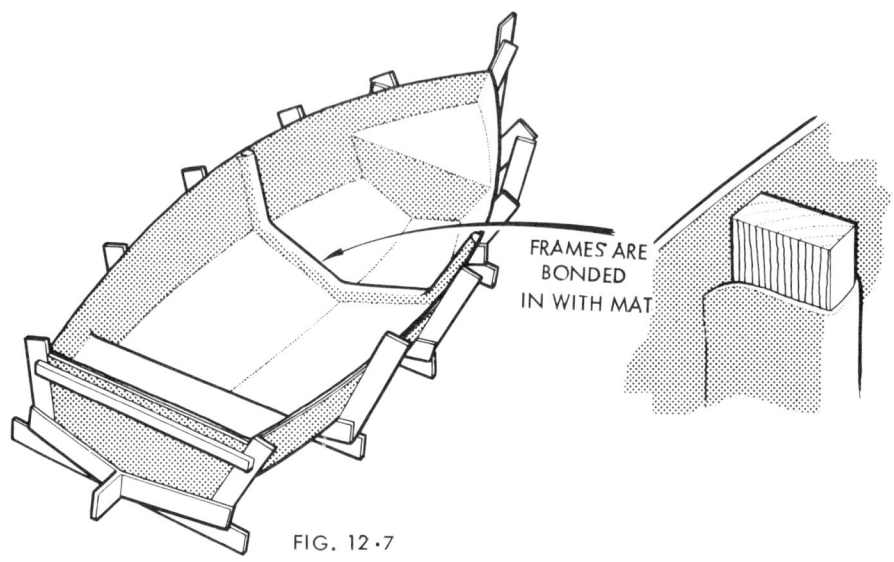

FRAMES ARE
BONDED
IN WITH MAT

FIG. 12·7

Interior frames, etc., are normally constructed of solid mahogany, or sometimes ply, and bonded in place with strips of glass mat, as shown in *Fig. 12.7*. These are fitted after the moulding has set, but before it is removed from the mould. There is then little or no risk of distorting the hull shape.

Decks can be of marine ply, or separate GRP mouldings. Logically a GRP hull should have a GRP deck, but for amateur construction at least a ply deck is much simpler to cut and fit. A separate female mould is needed to produce a GRP deck moulding. The same applies where the boat is to have a cabin as well. Ply and solid mahogany construction will provide the quickest, simplest and cheapest answer for a one-off job. If a GRP moulding is considered, however, this will have the advantage that deck and cabin sides and roof can be moulded in one piece and will give a perm-anent structure free from leaks, if properly fitted.

*Lay-up Recommendations*

Glass mat is to be preferred to glass cloth for GRP hulls, mainly because it is easier to ensure complete and thorough wetting with resin using this material. Hulls laminated from cloth have, in the past, often suffered from de-lamination, where the first or out-side layer starts to peel off, parti-cularly if the workmanship was not up to suitable standards in the first place.

Glass mat can therefore be taken as suitable for all types and sizes of boat hulls, provided it is used in suitable weight or total thickness. If additional strength is required for any particular reason, then glass mat may be inter-posed with a layer of woven rovings.

Actual thicknesses used vary con-siderably with different designers. Those intent on economy of materials will use minimum thicknesses. Thus one finds recommendations of two layers of $1\frac{1}{2}$ ounce mat as suitable for hulls up to 10 ft long or more. In practice this must be considered inadequate as although such a hull may be strong enough it will inherently lack rigidity and be very springy. Also, apart from reducing cost, there is nothing to be gained by using a minimum thickness of hull moulding. GRP construction is very light, and GRP hulls are often so light as to need additional ballast. It is more realistic to incorporate more bulk and weight of GRP in the actual hull.

The following recommendations are given as a general guide, based on the use of glass fibre mat reinforcement throughout:

*Hulls up to 10 ft*—$4\frac{1}{2}$ ounces of glass per sq. ft—i.e. effectively three layers of $1\frac{1}{2}$ ounce mat.

*Hulls 10 to 14 ft*—6 ounces of glass per sq. ft—i.e. four layers of $1\frac{1}{2}$ ounce mat or three layers of 2 ounce mat.

*Hulls 15–18 ft*—$7\frac{1}{2}$ ounces of glass per sq. ft—i.e. five layers of $1\frac{1}{2}$ ounce mat.

*Hulls 18–24 ft*—9 to 12 ounces of glass per sq. ft, with a minimum of five layers.

These thicknesses apply overall. On certain designs, notably fast run-abouts and speedboats, and keel boats, thickness will need increasing locally with additional layers of mat. Thus an 18-ft speedboat, for example, may have a hull based on 5 layers of $1\frac{1}{2}$ ounce mat, but with up to possible 15 layers in highly stressed points, such as around the chines. The transom, too, will need additional stiffening, especially if it is to take an outboard motor or an inboard-outboard unit. Virtually standard practice here is to use sandwich construction for the transom, with the core of marine ply ranging in thickness from $\frac{3}{4}$ in. minimum, upwards.

Where weight saving is important, as in the case of a racing yacht, thickness may vary from a maximum at the keel to a minimum over the topsides. For example, as little as 3

ounces per sq. ft may be considered adequate weight for the hull above the waterline, increasing to $4\frac{1}{2}$ ounces per sq. ft in the deck areas, and below the waterline. In the case of a keel yacht this would be further increased in the region of the keel, depending on the keel weight.

For readers wishing to study this subject in more detail, a copy of Lloyd's recommendations for GRP hull construction will provide a useful reference. It should be pointed out, however, that most professional builders work to their own specific ideas, and usually in excess of Lloyd's requirements.

The lay-up of glass fibre hulls follows the same technique described in Chapter 7. The glass reinforcement used should be of "E" grade (see Chapter 2). Surfacing tissue or surfacing mat is recommended for use both immediately following the gel coat, and as a final layer on the lay-up, although this is not invariably employed. Thixotropic resins are recommended throughout to eliminate draining. Any colouring or fillers used should be restricted to the gel coat.

*Sheathing Existing Hulls*

Glass fibre is often widely recommended for sheathing existing wooden hulls. It can be applied to new hulls to improve their durability and water-tightness; or to old hulls to renovate them and give them a new lease of life. Its use in such applications is almost entirely confined to amateur work. At one period some new hulls in plywood construction were sheathed in glass fibre. Today when it is thought desirable to sheath a new ply hull the sheathing material used is nearly always nylon.

Glass fibre sheathing as a means of renovating an old wooden hull has many apparent attractions. Two or three layers of mat will build up a skin of sufficient thickness to provide

FIG. 12·8

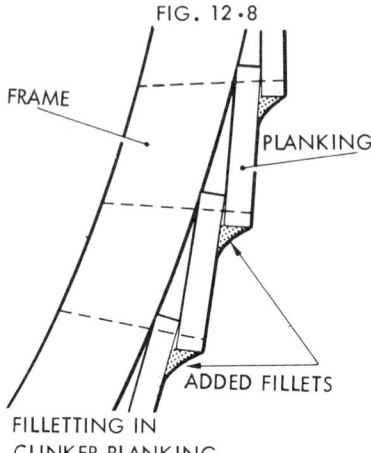

FRAME

PLANKING

ADDED FILLETS

FILLETTING IN
CLINKER PLANKING

complete strength, yet at the same time retain sufficient flexibility for the hull to "work", if necessary (this applying more to planked hulls than ply skinned hulls). It will also give the wood a permanent waterproof coating on the outside.

In practice, these advantages can be largely offset by the enormous amount of time and effort which may be needed to bring the original hull into a suitable state *for* sheathing. Also unless the work can be carried out in dry conditions, no attempt at sheathing will be successful. The original hull wood must be *thoroughly dry* throughout, which may take several months even under favourable indoor storage.

Drying out of an old clinker or carvel built hull will also result in shrinkage, although this need not present a major problem. In the case of carvel construction, all loose caulking should be removed (clinker hulls are not caulked), and made good by filling flush with the surface with resin/filler or polyester filler. Original defects made good with putty or marine stopper must be gouged out and refilled with resin/filler.

Before any of this can be started, however, the whole hull surface must be stripped right down to bare wood. Paint is best removed with a paint stripper rather than burning off, as charred wood will not provide a satisfactory bond for the resin and may inhibit setting in such areas. Thus all paint, and all oil-bound fillers must be removed.

It will also be necessary to remove any grease or oil which may have been absorbed by the wood. Swabbing the affected area with carbon tetra-chloride or strong detergent solution is probably most effective here, but again may take considerable time to get all the oil out of the wood. Naturally oily wood, like genuine teak, may not be suitable for sheathing at all. It largely depends on how "dry" it has become over the years.

All traces of rot or damage must also be cut out and repaired with new wood, or polyester filler. In the case of a clinker built hull it will also be advisable to fair in the bottom of each plank with lengths of triangular section wood, as shown in *Fig. 12.8*, to eliminate the sharp edges which would otherwise be difficult to cover round satisfactorily with the sheathing. These wood strips should be glued in place, using conventional resins, and also nailed at intervals to hold in place until set. Only copper nails should be used for this purpose, which will require the drilling of a hole to take each nail.

The success of a sheathing job will be as good as the preparation of the outer hull surface to a smooth, clean, paint and grease free condition—and whether the wood is really dry or not when the actual sheathing is done. The sheathing part is usually the least laborious part of the whole job; but it will be the most expensive, for a considerable quantity of glass mat and resin may be involved.

For a hull which is reasonably sound and does not require sealing of major leaks, a single layer of 1½ ounce or 2 ounce mat may be adequate, applied over a gel coat and finished off with a second gel coat or surface tissue. For hulls in poor condition two or three layers of 1½ ounce may be needed. Thixotropic resins should be used throughout. Note that sheathing should only be done on the outside of the hull, never to both sides.

Fillers should not be used in the resin. Pigments may be added for colouring, although again plain resin is probably best using an overall paint finish for colouring. To finish for painting, the sheathing can be sanded down with very fine abrasive after the resin has set hard. It should then be given two or three coats of polyester varnish or polyurethane varnish, to which pigment can be added for colouring. Normal paint finishes will not adhere properly to glass fibre, unless the surface is first treated with a suitable etching primer. In any case, polyester or polyurethane varnish will provide a more durable and more water resistant coating than ordinary marine paints.

*Deck Sheathing*

Glass fibre scrim applied directly to wooden decks with polyester resin provides a durable, waterproof covering far superior to canvas. The resin used can be pigmented to give a suitably coloured deck. Using a single layer of reinforcement the scrim pattern will be duplicated in the surface, if the resin coating is not too generous, which will impart a proportion of "non-slip" properties. This can be improved by dusting the resin with fine sand, whilst still wet and tacky, and smoothing out uniformly. Loose sand can then be brushed off when the resin has set.

Similar considerations apply as for sheathing. The deck must be stripped completely of old paint, etc., and cleaned right down to bare wood to ensure proper adhesion of the resin. If the deck has previously been covered with a rubber sheeting material, a solvent should be used to remove any remaining traces of adhesive. Above all, the deck surface must be dry before attempting to apply the resin and scrim.

# 13.  CAR BODIES

IT is doubtful these days whether the home construction of GRP mouldings to replace damaged car bonnets or wings, etc., is a worthwhile proposition owing to the ready availability of commercially produced mouldings for most of the popular models. A damaged body unit, even if it can be removed intact, may need a considerable amount of work to re-build to a suitable shape to use as a pattern, from which a GRP female mould must then be taken before a final moulding can be produced. Ignoring the work involved, the cost alone may well approach that of a ready-made GRP moulding.

The same comment applies to some extent to complete car bodies for "specials". Again such bodies are available commercially, although the cost in this case can be relatively high. The "special" builder may, therefore, be tempted to build his own GRP body, starting from scratch. The production of a one-off car body from scratch will, however, involve a considerable amount of work before the actual body can be moulded. The following basic stages are involved.

*Design*

The critical dimensions involved in design are summarised in *Fig. 13.1*. This basic layout should be drawn up, to a suitably large scale—not smaller than one-tenth full size, and preferably larger.

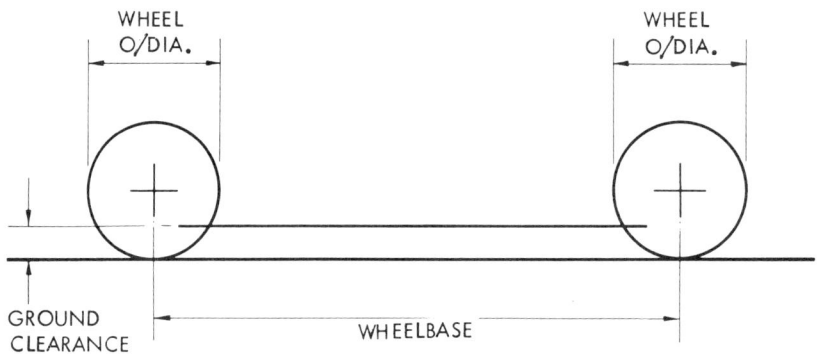

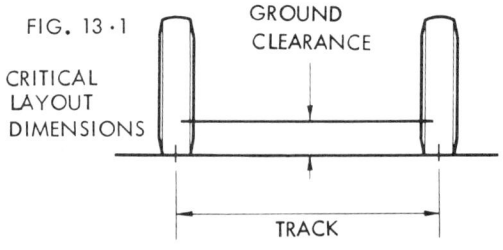

FIG. 13·1

CRITICAL
LAYOUT
DIMENSIONS

FIG. 13·2 BASIC LAYOUT

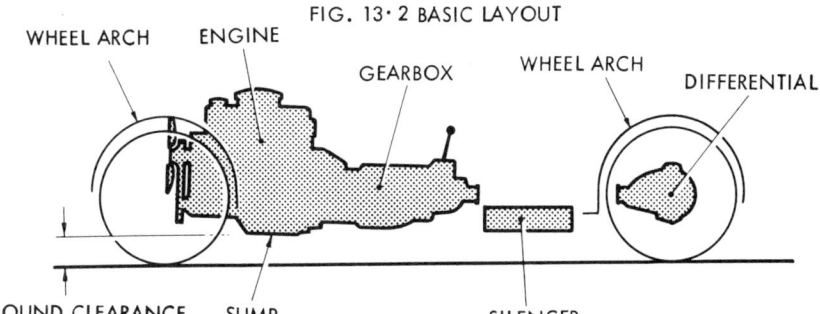

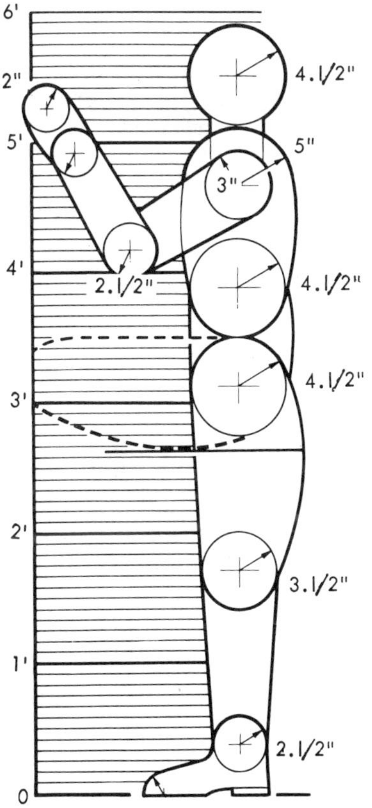

WHEEL ARCH  ENGINE  GEARBOX  WHEEL ARCH  DIFFERENTIAL

GROUND CLEARANCE  SUMP  SILENCER

6'

2" 4.1/2"

5' 5"

3"

4' 2.1/2" 4.1/2"

4.1/2"

3'

2'

3.1/2"

1'

2.1/2"

0

HUMAN FIGURE PROPORTIONS

FIG. 13·3

The *wheelbase dimension* largely determines the disposition of the "working" components—engine and gearbox, propeller shaft and differential; and the *ground clearance* the lower level limit for the engine sump and body floor, bearing in mind that appendages such as the silencer must be taken into account. The *wheel sizes* can then be drawn into scale to give the required position and clearances for the wheel arches (*Fig. 13.2*).

Provision now has to be made to accommodate the driver and passenger(s) in the overall design. A scale human figure, with correct joint pivot positions is shown in *Fig. 13.3*. These proportions are typical for a 5 ft 10 in. man. They can be adjusted to other heights, if necessary, bearing in mind that difference in height in human figures is *usually produced by differences in leg length* only, not overall scaling up or down.

A simple pattern of this figure can be constructed in thick card, with joints pivoted with paper fasteners, to the correct scale. This can be positioned over the drawing to determine seat position and driving attitude; and also the best control positions. Once the figure has been adjusted to a suitable position, it can be drawn round (*Fig. 13.4*). The side view body shape can then be drawn in to suitably envelop all the "contents" involved.

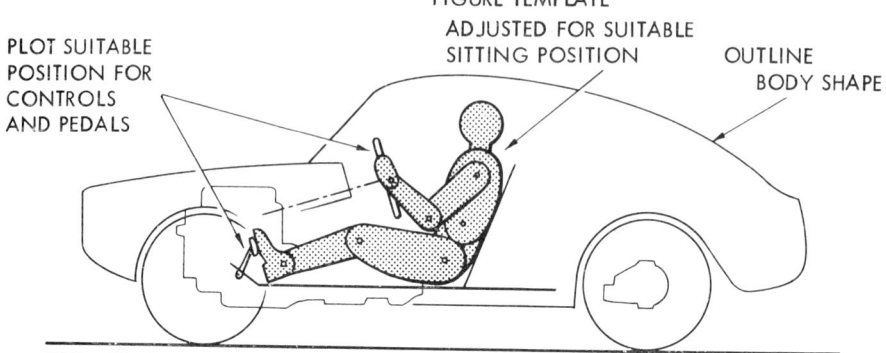

PLOT SUITABLE
POSITION FOR
CONTROLS
AND PEDALS

FIGURE TEMPLATE
ADJUSTED FOR SUITABLE
SITTING POSITION

OUTLINE
BODY SHAPE

FIG. 13·4 POSITIONING THE DRIVER

An end view drawing is then prepared. The critical dimension in this case is the track (*Fig. 13.5*). Bear in mind the body width to be accommodated, which determines the width of the seats, and the necessary clearances across the section.

### Scale Model

A scale model should then be made from these drawings. This can be a relatively simple mock-up, based on wood blocks, etc., finished off with plasticine or modelling clay, or carved to shape. The former is the more flexible method, for it enables detail changes to be made very easily, and is also much quicker.

### The Plug

A full size mock-up or plug must be constructed next, scaled up from

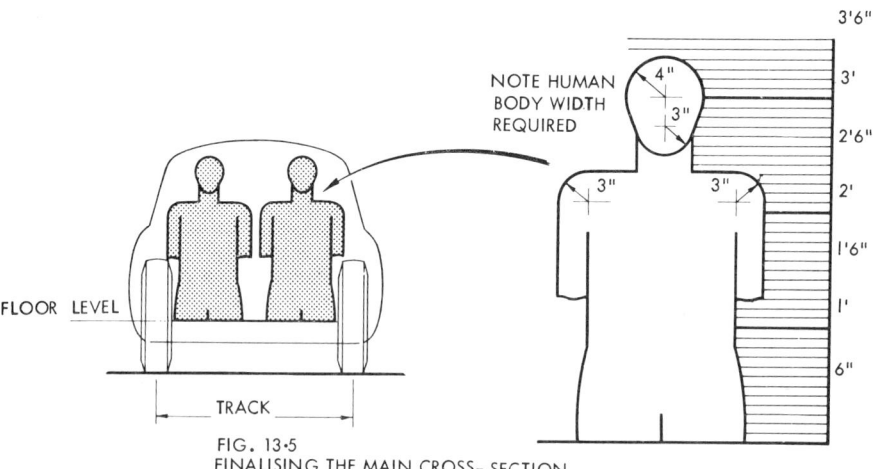

NOTE HUMAN
BODY WIDTH
REQUIRED

FLOOR LEVEL

TRACK

FIG. 13·5
FINALISING THE MAIN CROSS- SECTION

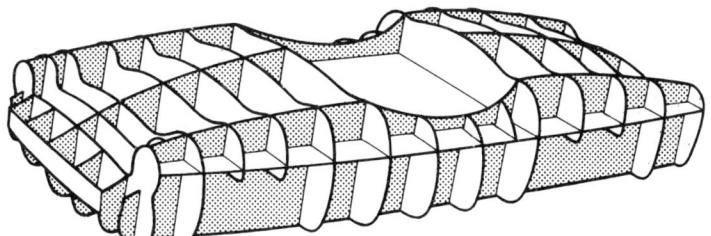

FIG. 13·6 RGG BOX CONSTRUCTIONAL FOR BODY PATTERN

the model. Again slight changes can be introduced, if thought necessary. Equally, the scale model stage *can* be omitted entirely, relying on introducing necessary changes from the original drawing in the full size mock-up stage. However, a scale model does give a "preview" of the mock up, and is a worthwhile extra stage. Much depends also on whether the builder is more competent as a draughtsman or modeller. In the former case he will probably get better results by going direct from drawing to mock-up stage. In the latter case, the drawing will give necessary proportions only, and virtually all the styling and shaping will be done "by eye" through the model and mock-up stages.

The mock-up construction follows the same principles as that for making up any large mould—the simplest method usually being that of making a basic frame which can be draped with wire mesh, followed by hessian and plaster—see Chapter 5. Alternatively an "egg-box" construction can be used, to be draped and covered with plaster—e.g. *Fig. 13.6.*

### The Female Mould

Whilst a moulding can be taken directly off the mock-up, this will have a rough outer surface. In view of the amount of work already done by this stage, therefore, it is far better to lay up a female mould in GRP over the mock-up from which the final body moulding is taken. Although this involves additional time and expense, the finished result will more than justify this difference. There is also one positive saving in that the moulding taken off a female mould can be self coloured, eliminating the rubbing down and painting which would otherwise be necessary with a moulding taken directly off the mock-up.

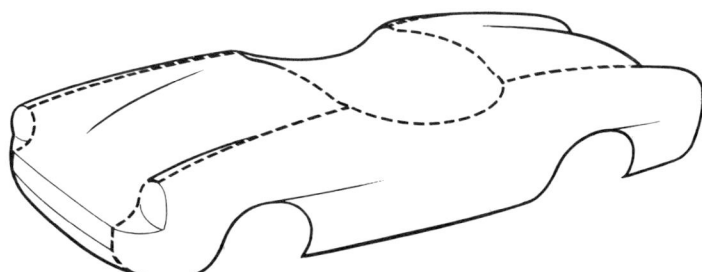

FIG. 13-7    EDGE TRIMMING LINES AND SEPARATION LINES
FOR BONNET

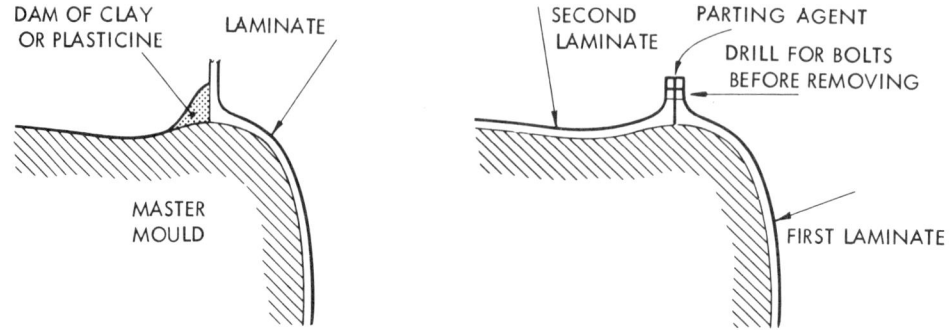

FIG. 13·8   INCORPORATION FLANGES ON SEPARATE MOULDINGS

First the mock-up must be studied to decide how many separate parts are required in the final body shape. These should be reduced to a minimum, e.g. separate bonnet and possibly separate boot lid. Having decided on suitable joint lines, these are marked on the mock-up (*Fig. 13.7*).

The female mould is then laid up in stages, erecting a suitable dam or barrier along the joint line(s). This can be built up with plasticine, as shown in *Fig. 13.8*. The first laminate is then laid up to this line, carried up the edge of the dam to form a flange. When set, the original dam can be removed and the flange on the first moulding can act as a dam for the second moulding, suitably protected with parting agent. The female mould is thus built up in separate sections. Necessary external bracing should

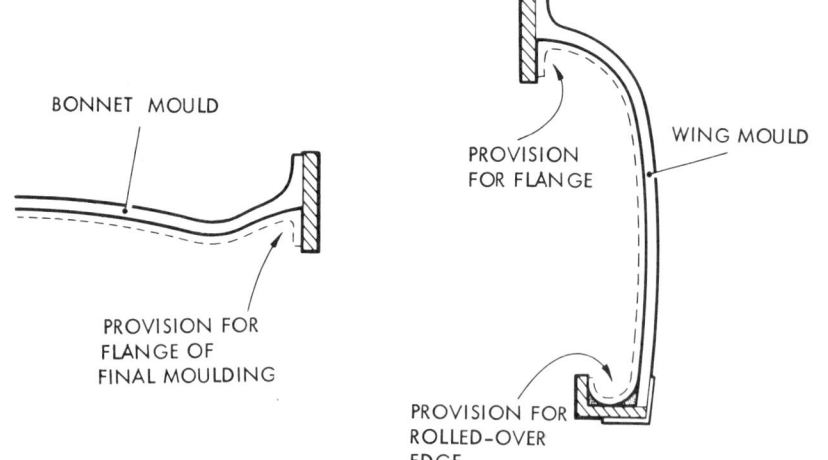

FIG. 13·9   MODIFICATIONS TO MOULDS

then be added to each section before it is removed from the mock up. Each section can then be removed when set, and further modified, as necessary, to provide flanges, etc., on the final moulding to provide rigidity at joints lines—e.g. see *Fig. 13.9.*

*Final Mouldings*

Making the final mouldings follows basic GRP lay-up technique. Generally two layers of 1½ ounce mat will provide adequate strength, but three layers may be preferred. Additional local reinforcement may be necessary in certain areas. Stiffeners may also be required, bonded to the underside of individual panels or sections. Attachment points for fitting to the chassis, local strengthening to take seats, etc., can also be bonded in whilst the moulding is still in the mould. A satisfactory ageing time must also be allowed in the mould so that there is no chance of the final mouldings distorting when removed from the mould.

## CAR BODY REPAIRS

This is one of the most common uses for GRP and resin/fillers. In fact, patching with GRP, and filling in dents, has largely replaced the traditional repair methods of cutting and welding in new metal panels, and panel beating, even with professional repairs. The process is basically straightforward, but must be done with due attention to detail requirements if a "clean" repair job is required, and the treatment is to be permanent.

Simple dents, scores and deep scratches are easily treated with resin/filler. The main requirement is to ensure that the area treated, and to which the resin/filler must adhere, is clean and durable. Ideally the whole area should be wire brushed down to bare metal to ensure maximum adhesion. Resin/filler can be applied over paintwork (provided the surface

has been degreased and de-waxed), but adhesion will then only be as good as that of the paint on the underlying body. Wax polishes are an obvious enemy of adhesion for they are release agents. Silicone waxes can also inhibit setting of the resin.

Resin/filler mix should be trowelled in place over a suitably prepared area, building up above the true body level. This is to allow for subsequent contraction on setting. If the dent is very deep, it may be best to fill in two or three stages. Flatting down can be done towards the end of the hardening time, before the resin has reached its full hardness, using fine abrasive. Full details of how best to use resin/body fillers for car repair work are normally supplied with car repair kits and so do not need further description in detail here. It may be mentioned, however, that most body fillers are based on a hard-setting resin which, due to the high proportion of filler used, may be relatively brittle when set. This can result in cracking under vibrational or twisting loads on the repaired panel. There may be some advantages in using an "elastic" filler in such places. An "elastic" filler (based on a plasticised or more flexible resin) is also much easier to flat down.

*Repairs to Rust Damage*

One of the most common repairs called for is to the bottom of doors or body sills where the metal has been eaten away with rust—and any remaining metal is frequently paperthin. The first step in the case of doors is to remove the interior lining panels (usually attached with screws) so that the inside of the door panel itself is exposed. The whole area to be treated then needs brushing over thoroughly with a wire brush to remove all loose rust and loose corrosion products. It is not necessary to clean right down to a bright metal surface.

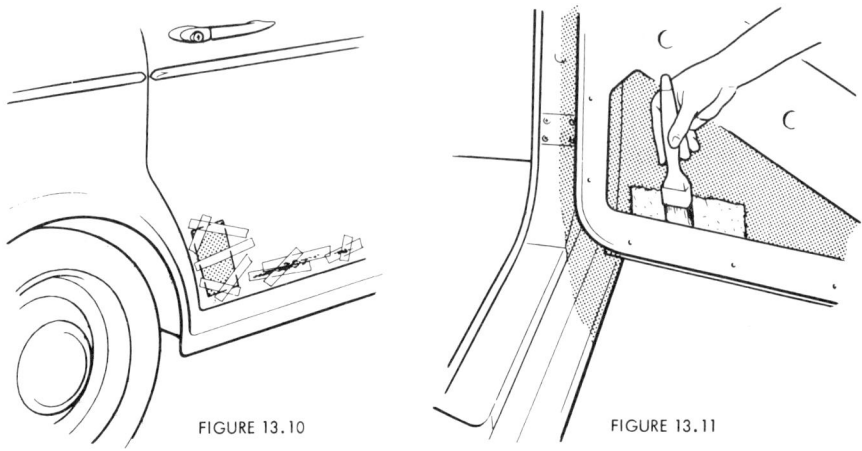

FIGURE 13.10                    FIGURE 13.11

The outside of the door panel should now be treated in a similar way, and any unsupported metal edges remaining bent back to shape.

If the area to be repaired is relatively large it must be covered on the outside with a rigid backing, such as hardboard, sheet balsa or similar material cut to conform to the shape required and capable of being lashed or clamped in close contact with the outer surface of the door. Before clamping up, a layer of thin acetate sheet should be placed between the door and the back-up material to act as a parting agent and give a nice smooth surface against which to lay the glass cloth or mat.

Small holes, etc., can be masked on the outside with Cellotape, simply using enough strips to cover all the holes completely to prevent the resin running through when applied from the inside—see *Fig. 13.10*.

Cloth or mat for the reinforcement is now cut to approximately the shape and size required, bearing in mind that this is to be laid in place on the inside surface of the door panel. Allow for two layers of cloth (or mat) as

giving suitable thickness and rigidity.

Now apply resin generously to the inside of the door panel. Lay the first layer of glass cloth in place and brush resin well into it, jabbing with the brush to make sure that the cloth lays flat over the repair area and also that it is completely impregnated with resin —see *Fig. 13.11*. The latter is indicated by the fact that the glass will turn almost completely transparent when fully saturated. The second layer of glass cloth or mat can then be applied right over the first, again coating with resin. The whole job can then be left to set for at least 24 hours before attempting to peel off the Cellotape or remove the "back-up" temporary structure.

The patch repair should now be quite hard and can be sanded down smooth prior to repainting to match the original colour of the door panel. If there is any roughness remaining, shallow spots can be filled with a resin/filler mix and allowed to harden before being sanded down. Small depressions are readily filled by "knifing" in a resin/filler mix. Exactly the same treatment can, in fact, be used

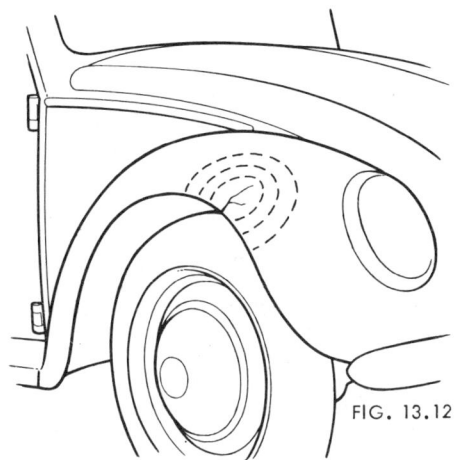

FIG. 13.12

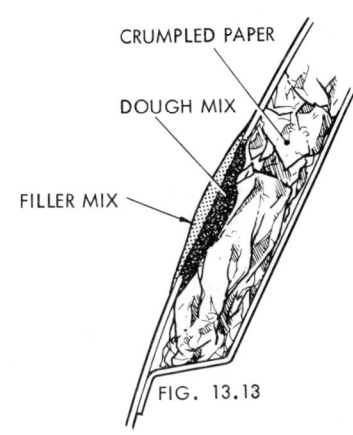

CRUMPLED PAPER

DOUGH MIX

FILLER MIX

FIG. 13.13

to fill in dents on bodywork, etc., instead of trying to knock them out. Simply level off with a resin/filler mix and sand down smooth when set.

*Repairs to Cracked Wings (Fig. 13.12)*

Repairs to cracked wings, etc., follow a similar technique in first brushing off rust, etc., from around the damaged area and then applying the resin/glass cloth "patch" from the inside. In this case, however, there is usually no need to back up the front surface, as by closing up the crack there is little or no seepage of resin through. Instead, two pieces of board —each faced with a piece of acetate sheet or polythene sheet—should be clamped up tight against the edge of the wing (on each side) to hold the two cracked portions together in their original position. This clamping up is done after the resin/cloth layers have been positioned. It will also be advisable to increase the number of laminations to three or four layers if the crack is at all extensive and the job needs holding strongly to get back to its original shape.

For applying the patches in really awkward places—e.g. right up inside a wing, or the top of a wheel arch where it may be difficult to get the layers to stay in place as you try to position them—the following technique is recommended. Lay a folded newspaper out over a flat surface and cover with a sheet of cellophane or polythene. Lay on a patch size of cloth and soak with resin. Add the second and further layers of glass cloth in the same way, impregnating each with resin in turn. The whole lot can now be picked up by putting one hand under the newspaper and simply placed in position over the patch area and smoothed and pressed into position. The whole patch area should, however, first be given a generous coat of resin which is allowed to gel off slightly to become tacky. Then the wet patch you picked up will have something to adhere to and will be easier to smooth out in position. If necessary the whole patch can be held in position (e.g. if it shows any signs of falling away) with suitable props.

Other useful techniques include the use of fine-wire mesh to fill in large gaping holes before applying resin/

cloth layers, or simply resin/filler mix. Since the resin adheres strongly to metals (provided they are clean and grease free) it can be used for sticking such metallic patches in place directly. A backing-up patch of cloth will be an advantage, if that side can be reached to apply it. Other areas requiring repair may be essentially "hollow" and not accessible from the inside surface to apply the necessary patch—see *Fig. 13.13.*

In this case a solution is to stuff the hollow interior out as far as possible with any suitable material—e.g. old newspapers crumpled up—and then apply a "dough" mix through the hole (see Chapter 9) to fill, as near as possible, to the original surface. Let the dough mix set and then make good and smooth up to surface level with resin/filler mix, finally sanding down smooth when hard.

In other cases it may be better practice to entirely re-make a panel or section, using the remains of the old part as a basic pattern or making a new pattern for the job. The replacement part in glass plastic can then be bonded in place with resin or, if more suitable, bolted in position to the basic structure or attached with self-tapping screws, rivets, etc.

Once having acquired a certain basic knowledge in the application of glass plastics and resin/filler mixes, suitable methods of approach to almost any repair or replacement job will automatically suggest themselves. There is almost no limit to the extent to which glass fibre can be used for car bodywork repairs, although it cannot be recommended for repairs to stressd metal structural members which have fractured or broken, or for blanking off splits or holes in silencers.

# 14. MODELLING APPLICATIONS

POTENTIAL applications of reinforced glass plastics are so numerous that it is virtually impossible to do more than describe a number of "basic" uses, leaving modellers themselves to exploit the almost endless possibilities with this versatile material One important point is that glass plastic construction is quite different from working with the more conventional modelling materials and thus, until familiar with handling it and the basic techniques involved, it may seem a more "difficult" material to use. No skilled model maker learned to shape wood or metals without practice, and the same is true with satisfactory glass plastic mouldings.

On the other hand it is a far more "foolproof" material in that shaping is controlled by a mould and laying up a laminate in a mould is a semi-skilled rather than a skilled job, so that one rapidly becomes familiar with the material and its characteristics. In other words, it is very easy to learn to use glass plastics properly, and extract full advantage from the material and the additional scope it brings.

Model boat hulls are a particularly logical subject for glass plastic mouldings, the exceptional strength/weight ratio being especially useful. A hard chine hull built up in the conventional fashion from ply and balsa skinning is a comparatively lengthy process,

FIG. 14·1

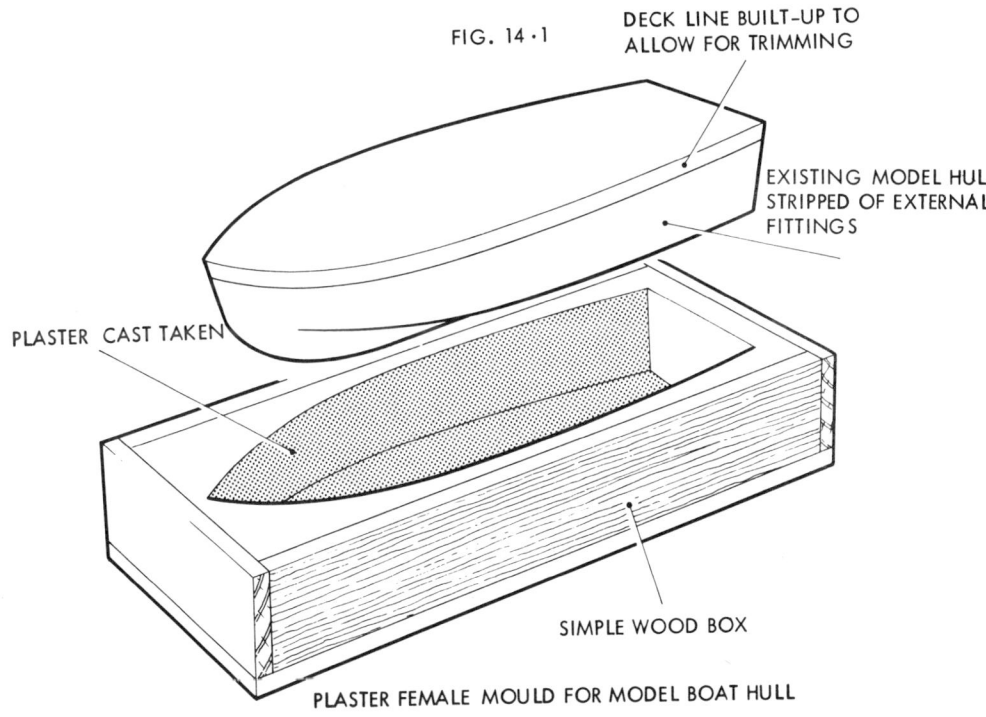

DECK LINE BUILT-UP TO ALLOW FOR TRIMMING

EXISTING MODEL HULL STRIPPED OF EXTERNAL FITTINGS

PLASTER CAST TAKEN

SIMPLE WOOD BOX

PLASTER FEMALE MOULD FOR MODEL BOAT HULL

and the watertightness of the resulting hull is not always as good as it might be—especially if cellulose cements are used with ply skinning. Round bilge hulls are usually avoided for model power boat use, since they have to be carved and hollowed out, whilst the carving or planking of a racing yacht hull is a tedious, laborious process. Apart from the fact that you still have to construct a mould of some sort, fabrication in glass plastic represents an easy and rapid solution.

For model boats up to about 10 in. in length, a single layer of 1½ ounce mat will provide more than adequate strength, but an additional layer may be considered advisable for rigidity. Two to three layers of 1½ ounce mat is adequate for hulls from 20 to 36 in. in length. For still larger hulls—up to 60 in.—three layers of 1½ ounce mat is recommended. Some modellers may prefer to use these recommended weights of glass but employ 1 ounce mat to increase the number of plies in the moulding.

The method of working with a female mould can be used to duplicate existing hulls. Thus a standard plastic or wooden hull can become a master pattern or plug, once exterior fittings have been removed and the gunwale line built up to provide extra depth for trimming. The female mould can then be produced in GRP, laid-up over the original hull; or cast in plaster from this hull (*Fig. 14.1*). The latter is a quick and simple method of producing female moulds for one-off jobs.

So popular have GRP hulls become for model boats that a wide range of proprietary mouldings are now available, covering both power boat and yacht designs. Unless particularly committed to an original design, the individual builder will usually find these a better proposition than making his own mould and moulding as the saving in time and effort is usually well worth the extra cost which may be involved.

Subsequent treatment of a moulded glass plastic hull is very much a matter of personal choice and ingenuity. By leaving a "lip" on the top of the moulding decks may be attached direct (*Fig. 14.2*), but this restricts the design to a straight deck line. It is more usual to trim the height of the moulding to the deck line required when the attachment of a wooden gunwale provides a fastening point for the deck beams as required, and the deck itself (*Fig. 14.3*).

Small glass plastic hulls—say up to 30 in. length—will normally not require any internal frames or stiffening at all, sufficient rigidity being given by the deck. Bulkheads may be fitted for convenience, or to blank off parts of the hull into watertight compartments. Equally, buoyancy compartments can be moulded in glass plastic and

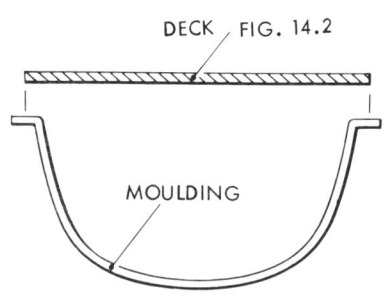

DECK　FIG. 14.2

MOULDING

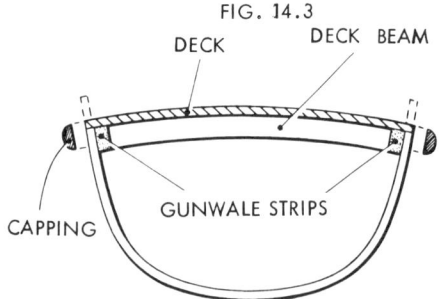

FIG. 14.3

DECK　　DECK BEAM

CAPPING　　GUNWALE STRIPS

added to the moulding to give an "unsinkable" hull. Frames and/or stiffeners will probably be required on larger hulls to ensure that the beam does not tend to spread, or even to control the final deck platform shape.

The absence—or near absence—of internal members, coupled with the thin skin required, means that a glass plastic hull is a very "roomy" hull—with plenty of space for installing the power plant, radio control gear, etc., if required. Such fittings as are required, like motor bearers, etc., can be inserted whilst the laminate lay-up is still tacky, or bonded on with strips of resin-soaked glass cloth later. The builder has a very wide freedom of design choice in this respect. The two important things to bear in mind are that the finished hull will only be as true as the original pattern and its strength and durability will depend on following the correct lay-up technique, eliminating air bubbles and ensuring adequate impregnation of the glass fibre reinforcement with the resin. Grade "E" cloth or mat should always be used for hull construction.

Other marine model items which can be tackled in glass plastic include complete superstructures with internal decks, deck cabins, etc., and even smaller fittings like dinghies, hatches, engine-box covers and so on. There is certainly no need to stop short at hulls when thinking in terms of glass plastic construction for marine models. You have at your call a material as strong as steel, easily formed to almost any shape (although very small mouldings are tricky to handle), and one which is light and permanent. Nor can the possibility of using resin castings be overlooked.

The strength of such a hull is, proportionately, very much higher than a full-size moulded glass plastic hull, with the weight still comparable—and probably less—than a built up or carved wooden hull. It can be expected to take almost any knocks the craft is likely to receive in service without damage. Using a pigment in the resin coats, the hull can be self-coloured and will continue to look smart and "new" for years. Alternatively, existing wooden hulls can be sheathed in glass plastic to give watertightness and added strength, as in the case of full size boat hulls—see Chapter 11. In such cases a single layer of 1 ounce mat should be adequate for the job on any size of hull.

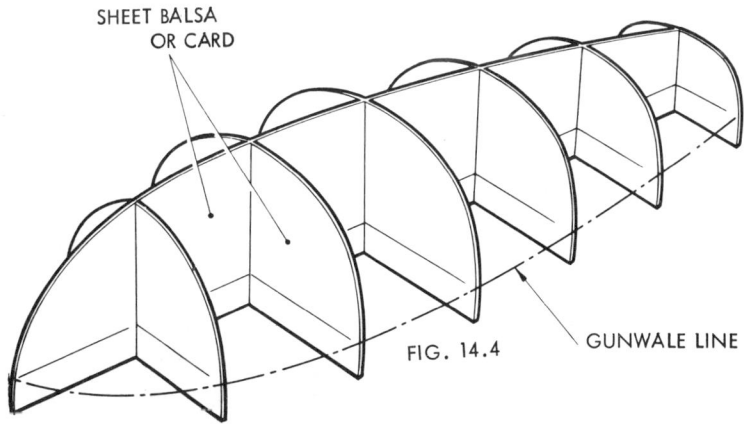

SHEET BALSA OR CARD

FIG. 14.4

GUNWALE LINE

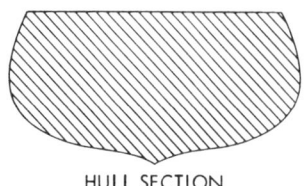

HULL SECTION

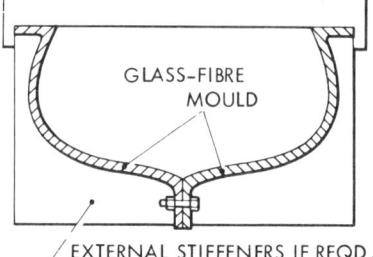

GLASS-FIBRE MOULD

/ EXTERNAL STIFFENERS IF REQD.

FIG. 14.5

A male hull mould can readily be built up from card or balsa patterns taken off the hull plan—assembling In skeleton form as in *Fig. 14.4*. The pattern can be completed by filling In between the bulkhead patterns with scrap material to bulk and completing the shape with plaster or plasticine. If smoothed right off (and in the case of the plaster mould, coated with shellac and waxed—see Chapter 7) this pattern can be used directly for making a moulding. However, this will produce a rough outer surface on the finished moulded hull and it would be better practice to use the pattern to prepare a female mould in glass plastic, and make the final hull mould-ing in this glass plastic mould. This may appear a duplication of effort, but the result is certainly worthwhile and the additional time spent not very great. The fibre glass female mould, too, is available for further use, if required, whereas the original male mould may have to be broken down to withdraw the first moulding, if the shape includes any undercuts. Under-cuts would be accommodated on a female mould by making this mould in two halves, to be bolted together as in *Fig. 14.5*.

An alternative method which can be used for making a hull which has marked undercutting or tumblehome is to eliminate the transom from the moulding. In other words, the stern end of the mould is left open. The moulding can then be sprung out of the mould without much difficulty and a separate transom then fitted to complete the hull—see *Fig. 14.6*.

Model aircraft represent rather different requirements in that although high strength is desirable, actual weight is the final criterion as far as free flight models go. Thus balsa wood is a standard structural material for model aircraft airframes. Neverthe-less successful free flight models have been produced from glass plastic mouldings—with shell mouldings for the fuselage and wing and tail panels. In the main, however, these are restricted to radio control design of generous size, where weight has a less critical effect on flight perform-ance, and now almost entirely confined to moulded fuselage construction. These are produced in the form of half shell mouldings, suitably jointed. GRP mouldings would not normally be considered for wings or tail surfaces, owing to the difficulty of constructing such components, and the excess weight, compared with foam plastic construction. The latter has adequate strength, at less weight, when skinned with thin balsa or hardwood veneer.

GRP is an excellent material for moulded cowlings. This part of a scale or semi-scale model is always difficult to produce. Close-fitting cowlings in balsa have to be hollowed out to a

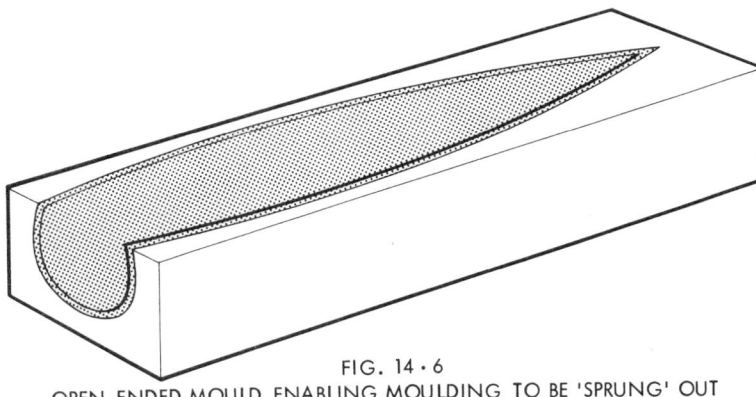

FIG. 14·6
OPEN-ENDED MOULD ENABLING MOULDING TO BE 'SPRUNG' OUT

degree where they are inherently weak —and the cowling usually comes in for a lot of handling in starting the engine. Glass plastic is an excellent alternative, and a far more satisfactory material.

The cowling shape required is best built as a simple balsa sheet mock-up, as in *Fig. 14.7*, when it can be filled in with plasticine to complete a solid form. A glass plastic laminate can be laid directly over this, or a glass plastic female mould made first so that the outer surface of the final moulding is smooth. A single layer of 10-thou. cloth should provide adequate strength for small cowlings, but an additional layer of 1-ounce mat (or cloth again) is generally to be preferred. Lugs or

suitable fittings for attachment can be mounted in the moulding whilst still wet, or bonded in later. Whether the cowling is designed to be detached and removed in one piece or split into two pieces depends on the model and motor mount it is to fit.

Spats, large fairings and similar relatively small, non-structural items can similarly be moulded in glass plastic. Another major use is for reinforcing the nose sections of large power models, particularly radio control models—see *Fig. 14.8*. A reinforcement of a single layer of thin glass cloth impregnated with resin is very much stronger than the conventional "bandage and balsa cement" reinforce-

PLASTER

FIG. 14.7

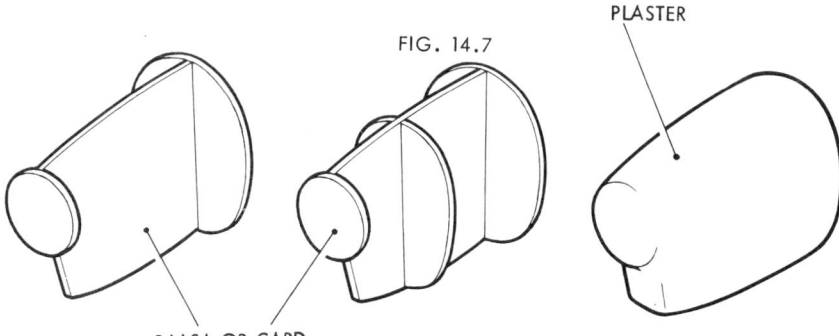

BALSA OR CARD

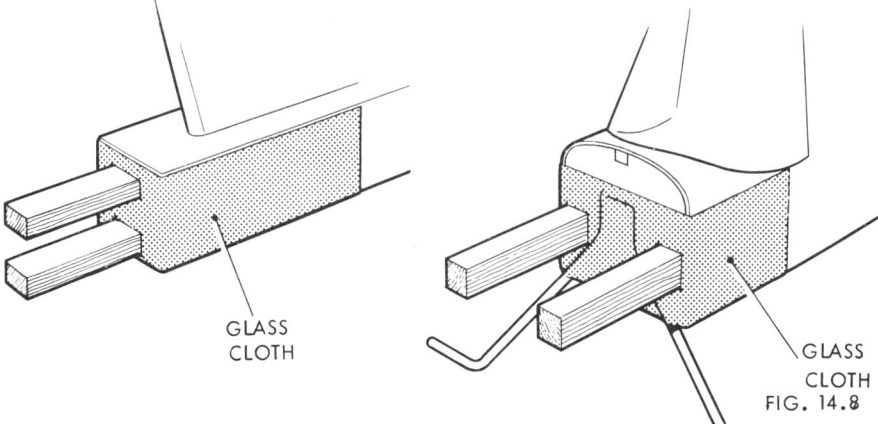

GLASS
CLOTH

GLASS
CLOTH
FIG. 14.8

ment. The same applies to local stress points on wing dihedral breaks and where wing attachment bands bear against the trailing edge (*Fig. 14.9*).

With control line models, weight is not so important and here the extra strength offered by glass plastics can be a considerable advantage. Stunt models are usually kept light, but with combat models (where toughness is essential), team racers and speed models there appears to be excellent scope for expanding the application of glass plastics—particularly as this field has hitherto been neglected.

The most logical application is to fuselages, moulded as half shells split either horizontally or vertically—see *Fig. 14.10*. A glass plastic pan can be made quite as strong as a light alloy "speed pan" of the same weight, whilst the upper shell can be kept quite thin and light as it is really only a fairing. The mould for the latter can be a simple mock-up in balsa and plasticine, and the time taken in producing a finished moulding probably no longer than doing a really first-class job in balsa construction. By making a female mould from the pattern and using pigmented resin, all the hard work associated with rubbing down and getting a good finish on a wood fuselage is avoided, and the problem of fuel proofing is overcome.

Similar considerations apply with team racer fuselages, although with a

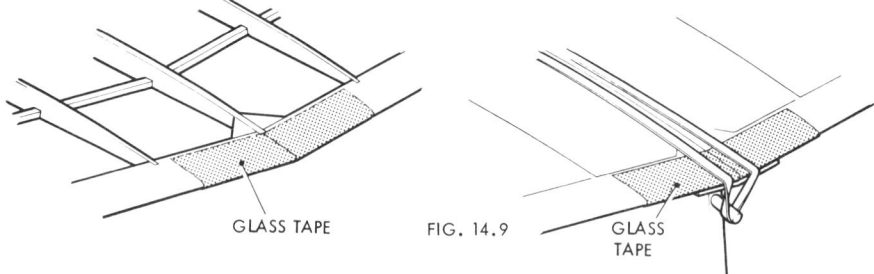

GLASS TAPE          FIG. 14.9          GLASS
TAPE

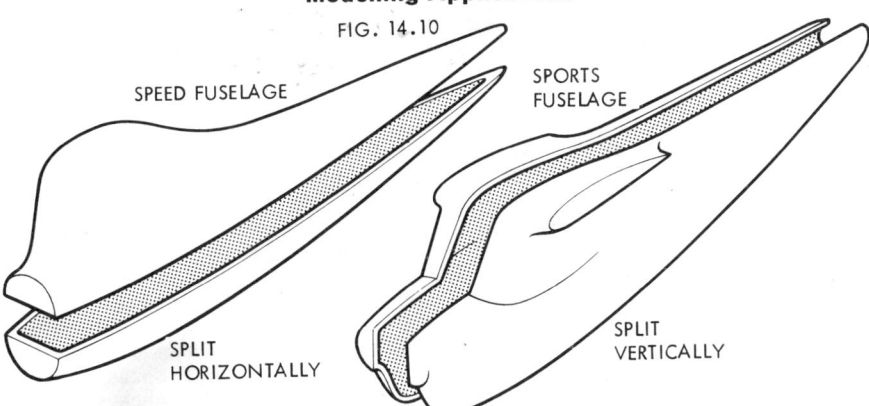

FIG. 14.10

SPEED FUSELAGE

SPORTS FUSELAGE

SPLIT HORIZONTALLY

SPLIT VERTICALLY

vertical split it will become necessary to incorporate internal bulkheads to carry the motor bearers. Once having produced a suitable mock-up fuselage and made a female mould from it, however, future production of team racer fuselages can proceed at a very high rate.

A glass plastic fuselage would probably best be associated with conventional wing and tail construction. Certainly there is no advantage in trying to make the tail surfaces from anything but balsa (or thin ply, if additional strength is required). The construction of wing "envelopes" or half shells in glass plastic hardly appears worth the extra trouble involved, compared with carving from solid balsa or using conventional built-up construction. Nor would "skinning" with thin glass cloth add greatly to the overall strength, the weakest point still being the bending strength at the roots where the wing joins the fuselage. Glass plastic sheathing would, however, protect leading edges against damage by "notching".

## Model Cars

For duplicating model car bodies rapidly for use on rail-track conversions, etc., a standard plastic car body can be a "ready made" pattern for making a female mould in glass plastic. From this mould any number of duplicates can be made with a minimum of trouble—in a variety of colours if you wish. Freelance car-body designs can be tackled on basic lines, starting with the construction of a full size plug.

## Model Railway Layouts

Model railway layouts can benefit greatly from the scenic effects such as hills, cuttings, tunnels and "landscape" forms, readily produced in glass cloth impregnated with resin and laid over chicken wire, balled-up newspaper, etc., to arrive at the required contours. This is probably the easiest method of tackling such layouts—very much quicker and simpler than using papier-mache, and more durable than hessian soaked in plaster.

The scope of glass plastic modelling can also be extended to architectural models, display layouts of all kinds and all other spheres where fabricated models are normally employed. Where certain items are duplicated (e.g. types of buildings), preparation of a single mould enables the complete number required to be turned out with the minimum of time and trouble.

# 15. DOMESTIC PROJECTS

GRP can be used to make a wide variety of domestic articles, furniture, fittings, etc. Many of these are simple variations on the basic forms already described (see Chapter 9), or suitable combinations of these forms. The scope is enormous, but rather than thinking of copying existing articles in GRP it is best to think in terms of designing an article to perform the same function but utilises to the full the advantages offered by GRP, notably the freedom of shape and the ease with which curved forms can be produced. This will give the GRP object character, which at the same time can often disguise some of the limitations of GRP mouldings (e.g. the fact that they normally have a rough and a smooth side).

The following notes summarise specific requirements.

## Draining Boards

With moulded-in ridges for draining, adequate stiffness should be given by two layers of 1½ ounce mat. Mouldings should be produced in a male mould similar to that for tray shapes (Chapter 9) using a coloured gel coat backed up by a layer of glass tissue. A sealing coat of polyurethane varnish is an advantage.

## Sinks

Produced on a male mould with sufficient draft for easy removal. Three layers of 1½ ounce mat should be adequate, with a coloured gel coat backed by surface tissue. Polyurethane varnish can be used for sealing both surfaces. Drain and tap fittings can be moulded in; or holes can be drilled out for the latter in the final moulding.

## Coal Hods

Usually best made in sections, bonded together, to which a separate hinged lid (tray shape) can be fitted. Three layers of 1½ ounce mat are recommended with additional edge reinforcement of tape. Hod mouldings should be through-coloured (i.e. using pigmented resin throughout). Colours may differ, e.g. a dark "decorative" colour for the gel coat, with black for the other colour. Shapes that can be used include parallel and straight tapered shells, and boxes with a separate "tray" end, if necessary.

## Garden Furniture

This is a classic example of where the curved forms and shapes readily produced by mouldings can be exploited to the full. Curved shapes will also add rigidity. The "top" face is always made the smooth surface on garden furniture, i.e. the fronts of chairs and the tops of tables. Leg sections can be moulded into the main shape, or legs can be attached separately—see *Fig. 15.1.*

Two to three layers of 1½ ounce mat is usually sufficient for all mouldings, although local stiffening may be needed. Through colouring is generally advisable, except for translucent panels used on tables. Alternatively the rough side of the finished moulding should be flock sprayed.

## Lampshades

Glass fibre lampshades are virtually a subject on their own. They are normally made from a single layer of 1 ounce or 1½ ounce mat, laid up around a suitable former such as a polythene bucket, plastic flowerpot, or similar article of suitable shape. Opaque and translucent colours can be used in the resin to eliminate the "neutral" appearance of glass fibre.

Very much more attractive lamp-tacking a layer of thin coloured material

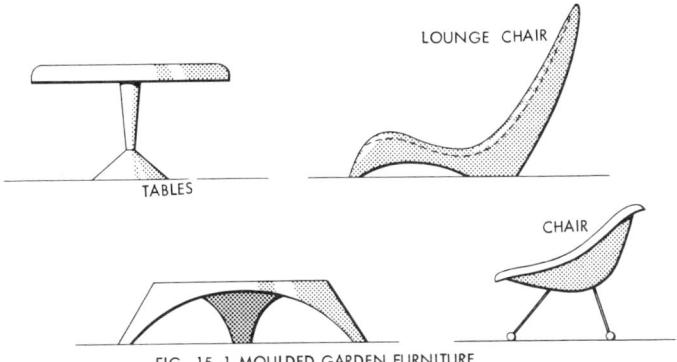

FIG. 15 ·1 MOULDED GARDEN FURNITURE

shades can be produced by first to the former, either plain or patterned. The glass fibre mat is then laid up directly over this, the two layers bonding together with the resin (*Fig. 15.2*).

Polythene containers are generally excellent for formers since they can be obtained in a wide variety of sizes and shapes. Also they do not need a parting agent to be applied before laying up the shade. With all other materials, a parting agent should be applied to the surface in the usual way, as for moulds.

Further possibilities include the embedding of leaves or similar flat objects with a decorative or artistic appeal in the resin lay-up (underneath the glass mat). These can be stuck to

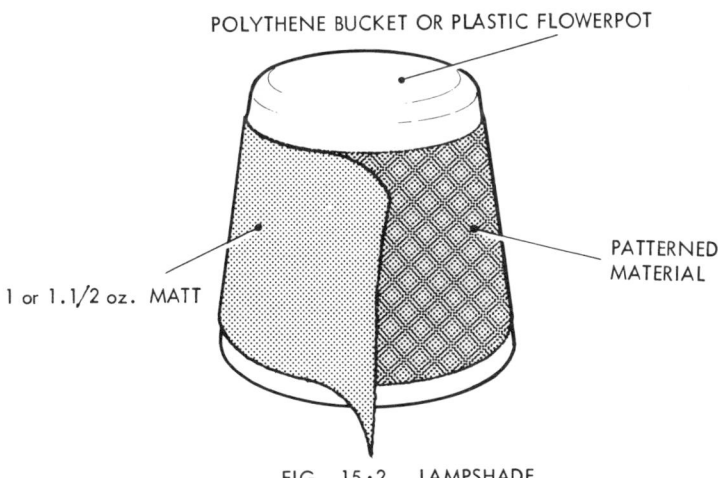

FIG. 15·2   LAMPSHADE

the former with dabs of PVA adhesive (white glue).

Fittings for the finished lampshades are usually best added by piercing around the rim of the moulding and sewing the wire in place. This joint is then covered with decorative tape or braiding, cemented in place.

## Household Repairs

*Repairs to Water Tanks*—see Chapter 11

*Repairs to Coal Hods, Buckets and Similar Metal Containers*

This follows the same procedure as for water tanks, applying the glass fibre reinforcement in the form of a layer on the inside, after first cleaning the surface as well as possible. In particular, remove any traces of coal dust, etc., from the surface of a coal hod, as this may inhibit the setting of the resin. If the article is to carry dry substances it may be used as soon as the glass plastic has set hard. If intended for carrying water or other liquids it must be left to set for at least 48 hours for the laminate to set really hard and assume its maximum resistance to water absorption.

*Repairs to Cisterns, etc.*

Highly satisfactory and permanent repairs have been carried out on cisterns which have cracked following a freeze up, even where whole sections have been broken away. The broken parts are replaced—they can be glued back with resin—and then reinforced with a glass cloth or mat layer, as in the case of tanks, etc.

*Repairs to Damaged Washbasins*

Whilst repairs can be made to cracked washbasins by sticking the parts back with the resin and reinforcing the damaged area with a layer of glass cloth and resin applied from the underside, adhesion to the highly glazed surface is not always as satisfactory as it might be. Where a piece has been knocked right out of the basin and lost, the underside patch repair can be made and the basin filled up to a level surface again on the inside with resin/filler mix. This type of repair, however, can only be regarded as a temporary remedy, although it will probably last for a considerable time.

*Repairs to Gutters and Downpipes*

Cracked or broken gutters, or those which have merely developed holes by corrosion are readily repaired with glass plastic. Preparation work required is to clean the inside of the gutter completely of loose corrosion products and make sure that it is perfectly dry. If necessary, use a blow-lamp to dry out the damaged area before applying the resin.

If the area of damage is extensive, a mock-up former must be fitted around the gutter to give the repaired area the required shape. This can be done with cardboard which has been given a coat of wax polish, then bent to shape and held in position with clamps, string, wire, etc., as appropriate—see *Fig. 15.2.* Two layers of 1- or 1½-ounce mat should provide adequate strength for a repair job of this nature.

In the case of broken downpipes, simply clean around the damaged area, coat generously with resin and lay on the glass mat (or cloth), stippling in position with a brush and more resin. When hard, the edges can be cleaned up by filing, if necessary, and the repair work painted over.

*Repairing Burst Pipes*

The best material for repairing burst pipes is usually glass tape or cloth cut to appropriate widths. The pipe surface must first be cleaned with a wire brush and dried, as necessary, using a blowlamp to speed the drying time. Then apply a coat of resin,

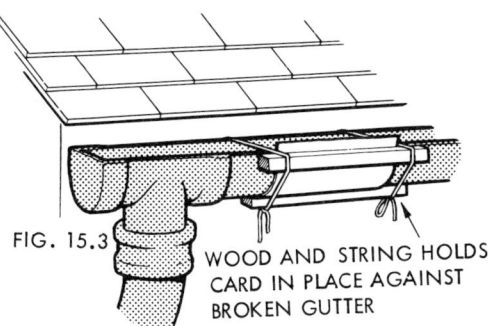

FIG. 15.3    WOOD AND STRING HOLDS
CARD IN PLACE AGAINST
BROKEN GUTTER

allowing at least an inch overlap each side of the crack, wind on the cloth strips or tape and thoroughly impregnate with a resin coat. Setting can be hastened by positioning an electric fire to warm the repair area, and must be complete—i.e. no surface tackiness apparent—before the water is turned on again and allowed to flow through the pipe.

This sort of repair is usually permanent as regards performance, but it is generally advisable to have a conventional repair made to the pipe later on, when convenient.

### Reinforcing Fence Posts

It has been found that the life of fence posts can be materially improved by sheathing the bottom ends with a glass plastic layer before they are driven into the ground. This usually provides better "proofing" than the usual method of tarring or coating with creosote, etc.

### Repairs to Leaky Roofs

Stubborn leaks in garden sheds, caravan roofs, boat cabin roofs, etc., can be sealed with a glass plastic "patch" over the appropriate area, or by filling with resin/filler mix, depending on the circumstances and nature of the fault. Again, of course, such repair work can only be applied when the area concerned is perfectly dried out. Patch repairs, too, should be applied on the *outside*. The main trouble in many cases here is in deciding exactly where the water is getting in and which is the "entry hole" that has to be sealed off.

### Repairing Holes and "Making Good"

An inexpensive "mix" for repairing holes, filling in cavities, etc., is a resin/sawdust mix compounded with the maximum amount of sawdust. This can be used for a wide variety of purposes where plaster, plastic wood, etc., might be considered as alternative materials and generally give a stronger bond and a better job. This type of mix can also be used effectively for filling in metal surfaces. In the case of large cavities to be filled, the bulk of the hollow space can be filled first with inexpensive bulk material, such as wood blocks, etc., and only the final surface repair done with the resin/filler mix.

# 16. GARDEN POOLS

GARDEN POOLS are one subject which really do not need the construction of a mould. Rather, the excavation for taking the pool can act as a mould. This gives considerable scope to the sizes and shapes possible, for one-off productions. The fact that the "finished" or top surface of the GRP pool will appear rough is no disadvantage. In fact, it is more natural in appearance than the commercially produced GRP pool liners which are almost invariably made on male moulds and thus have a smooth, shiny "finished" surface.

Certain basic rules apply in planning the proportions of a pool. These can be summarised as follows.

(i) Regardless of the actual size, no pool needs to be deeper than 30 in. This will be sufficient to ensure that the pool will not freeze solid in winter, and also be deep enough to accommodate plants such as water lilies.

(ii) A minimum depth of 15–18 in. is advisable, even for the smallest pools, in order to avoid excessive changes in water temperature between hot and cold weather.

(iii) Pond sides should slope outwards at an angle of about 20 degrees. This will eliminate the possibility of ice damage, and also ensure that the pool contains a reasonable volume of water for its surface area.

One other point can be considered before deciding on the pool proportions. Many aquatic plants, known as marginals, grow in fairly shallow water—preferring 3 to 5 in. of soil with perhaps 6 to 9 in. of water above. The design of a garden pool can therefore conveniently incorporate ledges around the edge on which marginal plants can be planted, or stood in pots. A ledge-shaped section will also be easier to excavate.

The basic requirements of a good pool design, therefore, work out as shown in *Figs. 16.1* and *16.2*. *Fig. 16.1* is for a *formal* or geometric shape which may be rectangular, square, or even circular or elliptic. *Fig. 16.2* is for an *informal* pool with an irregular outline shape. Having decided on a shape, suitable size, and site, the next step is to dig out the ground to this shape. This is the hardest job of the lot.

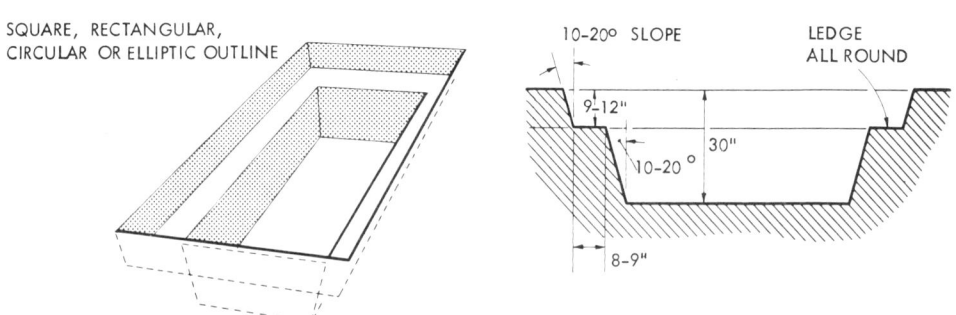

FIG. 16·1   FORMAL POOL PROPORTIONS

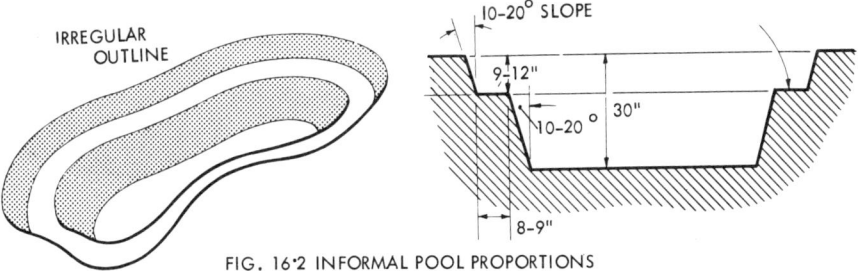

FIG. 16·2 INFORMAL POOL PROPORTIONS

It is important with a formal pool that the edges of the excavation should be level all round. If not the water level will lie at an angle to the edges of the pool and look wrong. This is less important in the case of an informal pool, but is again preferable. Check for level with a plank and a spirit level, laid across the excavation at various points (*Fig. 16.3*).

The sides and bottom of the excavation should be smoothed off and consolidated as far as possible. Sharp stones and other projections should be removed, and cavities filled in where necessary. All corners, and edges of shelves, should also be rounded off to simplify moulding.

The whole pool should then be lined with two or three layers of fairly stout brown paper. The object of this is to "hold" the resin when the gel coat is applied (in the absence of a proper mould surface), and also prevent contact between resin and possibly damp earth which could inhibit curing of the resin. For the same reason, all the work should be carried out on a warm, dry day and the paper used should itself be quite dry.

Laying up the moulding then follows normal GRP technique. First a gel coat is applied directly to the paper. Once this has gelled, glass mat layers can be added, with further resin, and stippled and rolled down (*Fig. 16.4*). The numbers of layers required depends on the size of pool. Two layers of 1½ ounce mat should be adequate for most pools up to about 100 sq. ft surface area, with additional layers for larger pools.

A final surfacing with glass fibre

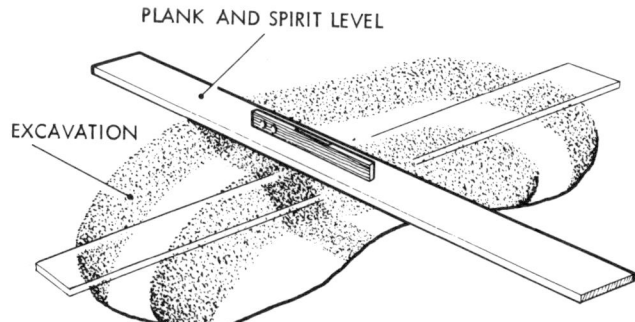

FIG. 16·3   CHECK EXCAVATION FOR LEVEL OF OUTLINE

(1)  SMOOTH OFF SIDES
      AND BOTTOM

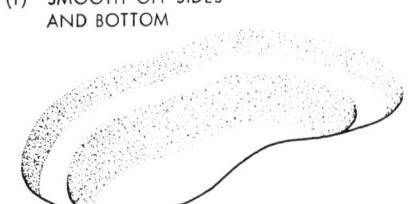

(2)  LINE WITH BROWN PAPER
      OVERLAPPING EDGE

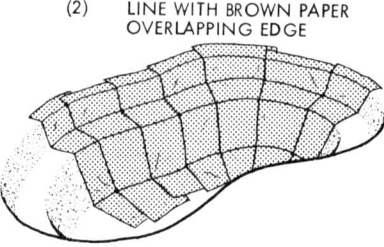

(3)  PAINT PAPER WITH RESIN

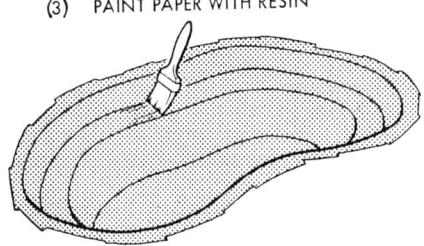

(4)  LAY UP GLASS MAT
      AND STIPPLE IN PLACE

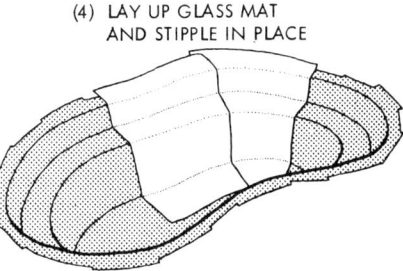

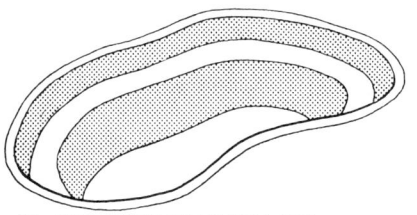

(6)  COVER OVER AND
      LEAVE FOR
      TWO WEEKS

(5)  TRIM OFF EDGE NEATLY WITH
      KNIFE, SCISSORS OR TINSNIPS
      BEFORE RESIN HAS FULLY HARDENED

FIG. 16·4     STAGES IN MAKING A G.R.P. POOL

tissue is usually worthwhile, although this may be omitted on the grounds of expense. In any case the finished moulding should be treated with a second gel coat on the "finished" side, followed by two or three coats of sealer or varnish to waterproof fully.

The finished moulding should also be left to cure for *at least* 14 days before filling with water. During this time it should be covered over so that it is protected against rain.

# 17. REPAIRS TO GRP

REPAIR work on GRP mouldings is readily tackled since polyester resin will adhere strongly to "set" and aged resin, or laminates, provided the surface is dry and greasefree. Repair procedure differs slightly depending on whether the damage is of minor or major nature, and where it occurs on the moulding.

## Surface Repairs

Damage to the gel coat may appear in the form of a fine network of cracks, or from scratches or gouges, etc., produced by wear. In extreme cases where the moulding has been badly abraded whole areas of the gel coat may be worn away, exposing the ends of "raw" glass fibres.

Gel coat crazing, or a network of cracks radiating from a single spot in the form of a "sunburst" may be the result of faulty lay-up technique (*Fig 17.1*). A more random distribution of cracks, forming a "cobweb" pattern is usually the result of impact (*Fig. 17.2*). In this case tap the moulding in the damaged area to check the laminate for soundness. If is sounds dull, crack damage may have extended through the laminate instead of being confined to the gel coat. Repair in this case would call for backing up the moulding with another "patch" layer or two of glass mat, as well as treating the cracks.

Fine cracks can be sealed by wiping over with activated resin. The crack pattern will still be visible, but the repair will effectively restore the seal originally provided by the undamaged gel coat.

Deeper cracks, or more pronounced cracks can be opened slightly by scratching with the point of a knife, or a finely pointed tool, taking care not to dig too deep. They should then be filled in with resin/filler or resin putty, left slightly proud. When set, this should be levelled off flat and the whole area polished to restore the gloss. Again the repair will still be visible, for an exact match of colour is virtually impossible. However, the thin line of slightly different colour will usually be inconspicuous.

Deep scores and gouges are treated in a similar way, except that there is no need to scratch them out. A brushing with a wire brush is, however, advisable, to remove any surface dirt which may have accumulated in the score.

Small scratches, caused by rubbing, etc., may only be superficial, in which case they can often be polished right out, using an abrasive cleaner, metal polish or a burnishing paste. Try this first to see if it works. Very fine scratches which cannot be removed by polishing out may respond to painting with acetone, but do not let excess acetone spill on to the main surface as it is a solvent for polyester resin.

FIG. 17·1  GEL COAT CRAZING  - EITHER FROM FAULT OF TECHNIQUE
OR RADIATING FROM A STRESS POINT

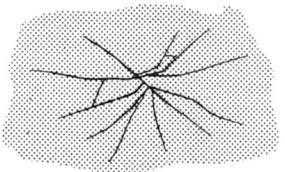 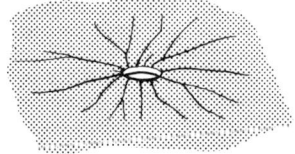

FIG. 17·2    'COBWEB' CRAZING
NORMALLY RESULTING
FROM IMPACT

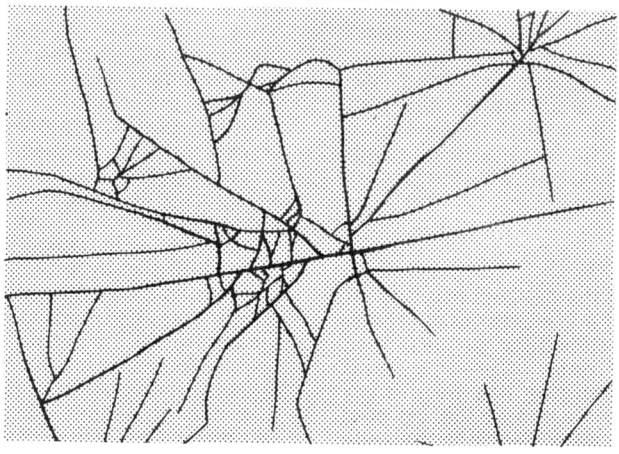

### Chipped Edges

These are more difficult to repair neatly. The damaged edge surfaces should be roughened by wire brushing, when a new edge can be built up, slightly oversize, with pigmented resin/filler. Simple shuttering, bent from card and covered with cellophane, may be required to hold the filler in place until set. This can be secured in place with adhesive tape. If a stronger repair is needed, use a dough mix instead of resin/filler or polyester putty.

### Deeper Cavities

Provided the damage has not completely pierced the thickness of the moulding, these can be filled with resin/filler or dough mix, as appropriate. The mix used should be pigmented to match the gel coat colour as closely as possible.

Inspect the moulding on the other side. If this shows signs of damage, back up the repair with a patch layer of 1½ ounce mat, applied with resin.

The same back-up treatment may be needed in the case of deep scores and gouges.

### Small Holes

When a moulding has been pierced right through the damage area will usually be further extended by cracks on the outer surface. The main job is to repair the principle damage first. Cracks can be filled in later, if necessary.

Cut the edges back in the form of a "V", as shown (*Fig. 17.3*). Any completely loose material should be cut away, even if this means enlarging the hole. The damaged area should then be backed up on the outside with a piece of hardboard or stiff card, depending whether the surface is flat or curved, faced with cellophane. This can be held in place with adhesive tape (*Fig. 17.4*).

Working from the other side, brush on a gel coat first. Follow this by laying up small pieces of 1½ ounce mat, torn to suitable size, and stippling

FIG. 17·3 PREPARATION OF HOLE DAMAGE FOR REPAIR

DAMAGE

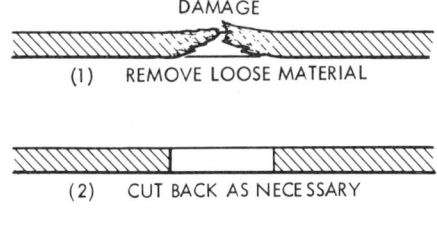

(1) REMOVE LOOSE MATERIAL

(2) CUT BACK AS NECESSARY

(3) CHAMFER EDGES OF HOLE

in place until the required thickness has been built up. If more than four layers are needed, stop at the fourth layer and allow to gel and harden before proceeding with the next layers.

Once completed properly a repair of this type should be as strong as the original material and will not need any back-up. The outer surface can be smoothed off and burnished, as necessary, to blend with the original surface, although colour matching will be

HARDBOARD OR CARD    CELLOPHANE    ADHESIVE TAPE

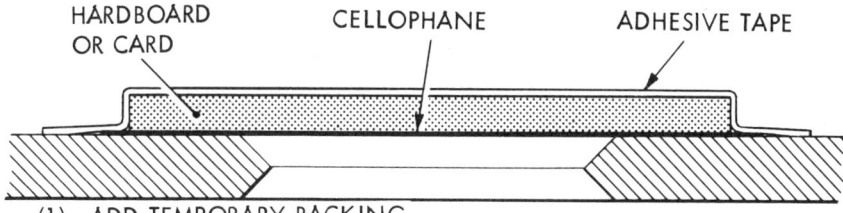

(1) ADD TEMPORARY BACKING

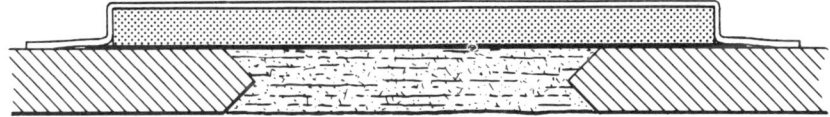

(2) FILL WITH LAYERS OF MAT TORN TO SHAPE AND IMPREGNATE WITH RESIN

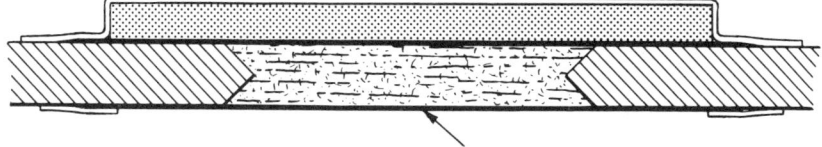

COVER UP WITH CELLOPHANE TAPED IN PLACE UNTIL RESIN HAS SET

FIG. 17·4 STAGES IN FILLING A SMALL HOLE

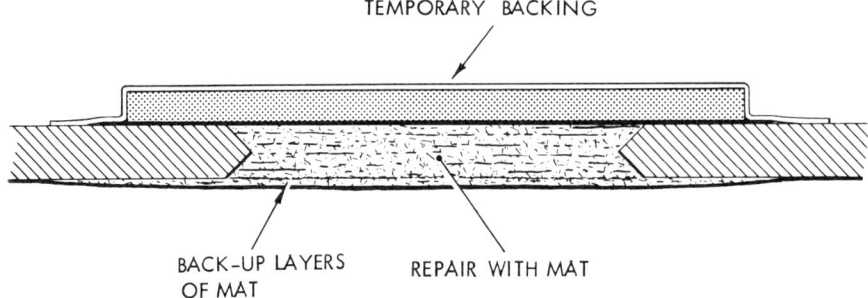

TEMPORARY BACKING

BACK-UP LAYERS OF MAT          REPAIR WITH MAT

FIG. 17·5   LARGE HOLE REPAIR

virtually impossible. It may be necessary to paint over to tone in with the original colour.

*Large Damaged Areas*

Trim right back to undamaged material, regardless of the size of hole opened up. This should include cutting out any obviously deep cracks, or cracks that run right through the thickness of the original moulding. Chamfer off the edges of the opening

FIG. 17·6 TREATING MAJOR DAMAGE   TRIM DAMAGED AREA TO REMOVE LOOSE PIECES

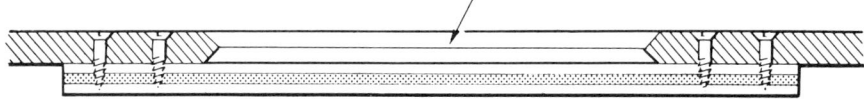

(1)   BACK UP WITH PLY SECURED WITH SCREWS OR BOLTS

(2)   COMPLETE REPAIR WITH MAT

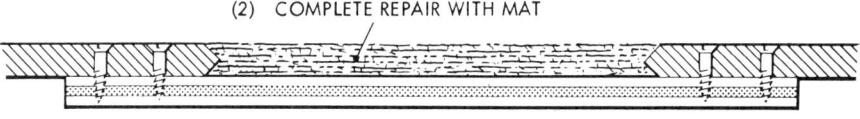

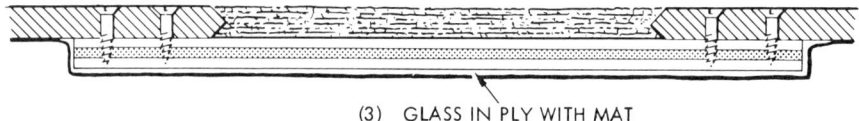

(3)   GLASS IN PLY WITH MAT

to a "V" shape and fit shuttering on one side (preferably the outside). Repair then proceeds as above, using larger pieces of torn mat to build up to the required thickness. A back-up layer or two of mat, extending several inches either side of the original damage, should also be added in this case (*Fig. 17.5*).

An alternative method sometimes used on boat hulls, especially in regions where an interior patch will not show, is to back up the damaged area with a large piece of marine ply, glassed in (*Fig. 17.6*). This arrangement can also be used for a repair which does not conveniently lend itself to temporary shuttering, or where the shuttering cannot be placed on the outside. In this case a piece of marine ply is used for shuttering on the inside, and permanently fixed with bolts or screws. The hole damage is then made good from the outside, finishing smooth by filing and sanding down when set. The shuttering is then glassed over on the inside to hold it in place, the original fastenings for the ply usually being left in place.

### Emergency Repairs to Boat Hulls

Emergency treatment may be necessary on a GRP boat hull to prevent water coming in through a damaged area. In this case the simplest, and obvious, treatment is to stuff rag, clothing, etc., into the hole to plug the leak; or if practical pull a sail or a polythene sheet against the hole from the outside. The latter treatment can be particularly effective since polythene clings to a wet surface.

If the hull has to be repaired in a hurry and put back in the water immediately afterwards, a *temporary* repair can be attempted with a wood or metal patch, bedded down with mastic or paint and secured with through fasteners (i.e. bolts). Self tapping screws can also be used, but are not as reliable as bolts. This patch repair should be stripped off as soon as it is practical to make a proper repair.

Ideally the patch should be put on from the outside, but if this is quite impractical, then it can be applied from the inside. In that case self tapping screws will have to be used.

Note some *epoxy resin* putties can be used under wet conditions, even underwater. They can be employed for repair work on scratches, etc., on a boat which is still afloat. Polyester resins and normal GRP formulations cannot, however, be used in very damp conditions, and certainly not underwater.

# 18.  TROUBLE SHOOTING

QUITE a number of things can go wrong with GRP mouldings, particularly in the hands of inexperienced operators. The main causes of faults are poor technique, working in damp conditions or extreme cold, or using the wrong proportions of materials. The following is a fairly detailed guide as to faults that can occur, and their causes.

## Gel Coat Faults

| Fault | Cause |
|---|---|
| Does not set | Lack of catalyst or lack of accelerator, or both.<br>Use of wrong type of catalyst.<br>Excessively low temperature.<br>Damp filler used (this can occur particularly with sawdust filler).<br>Excessive amount of pigment.<br>Use of a pigment or filler which has an inhibiting effect.<br>Damp mould. |
| Sets in patches, some areas not setting | Damp mould.<br>Damp brush used to apply resin.<br>Catalyst not uniformly distributed by stirring in (i.e. mix not stirred enough).<br>Presence of undried PVA release agent on mould surface, or in crevices. |
| Surface remains sticky | Not enough catalyst or accelerator (check proportions used).<br>Excessive evaporation of styrene (cover surface exposed to air with cellophane).<br>Damp mould.<br>Unsuitable parting agent.<br>Mould surface not sufficiently sealed. |
| Pinholes | Small air bubbles trapped in resin, probably caused by too vigorous stirring of resin mix.<br>Dust or dirt in resin or brush. |
| Wrinkles | Gel coat too thin.<br>Resin mix too slow in setting.<br>Draughts (cover mould, or remove from exposed position).<br>Application of another resin coat before the gel coat has gelled properly. |

| | |
|---|---|
| Blisters | Large air bubbles trapped under surface. Excessive exothermic heat due to too thick a lay-up being attempted in one go. |
| Glass pattern showing through | Use first layer or glass surfacing mat to hide glass pattern. |
| Patchy colour | Pigment not uniformly dispersed in gel coat resin. |
| Hairline cracks | Excessive exothermic heating. Excessive bending or flexing of moulding during removal from mould. Resin too brittle and not suitable for a gel coat. Sticking in mould, with damage arising when prised off. |

### Faults in the Complete Laminate

Troubles in the bulk of the laminate will be less visible since the appearance is judged by the surface. Hence faults present may often be identified as gel coat faults, as in the preceding table. The following specific faults may, however, appear:

| Fault | Cause | Remarks |
|---|---|---|
| Delamination | (i) Insufficient catalyst or accelerator<br><br>(ii) Faulty lay-up | It is possible to forget to add catalyst (or accelerator) when applying one layer<br>E.g. not wetted out properly |
| Sticky laminate | (i) As for "not set" above, in gel coat table | |
| Laminate sticky inside | (i) Damp filler<br>(ii) Excess of pigment<br>(iii) Lack of catalyst or accelerator | |
| Tacky surface, rest hard | (i) Mix too long in setting<br><br>(ii) Damp atmosphere<br>(iii) Draughts | Applicable to female mould production<br>A remedy is to wash off with acetone |
| Dry strands at top edges | (i) Drainage<br><br>(ii) High temperature | Use a thixotropic resin on vertical surfaces<br>Aggravates drainage |
| Sticking to mould | (i) Not enough parting agent | Wax and PVA may need to be used instead of a single parting agent |
| Porous laminate | (ii) Insufficient resin | Layers not wetted out properly |

# 19. DESIGN IN GLASS FIBRE

THE general properties which make GRP an attractive structural material are:

(i) High strength with low weight.
(ii) High impact strength combined with great resilience.
(iii) Suitability for making products of any size.
(iv) Resistance to weather, chemicals, water, most oils, spirits, acids, alkalis and solvents.
(v) Dimensionally stable—retains shape under mechanical stress and temperature extremes—from below freezing point to above 250°F.
(vi) Non-conductive, with good electrical properties, and ability to pass radio waves.
(vii) Low thermal conductivity; transmitting less than half the heat of glass.

(viii) Translucent naturally—may be coloured as desired.
(ix) Negligible water absorption with retention of shape and strength when constantly immersed.
(x) Easy to use without expensive equipment.

In the main design in GRP for stressing purposes is done on empirical lines. That virtually means "guesstimating" the thicknesses required, and whether or not additional local reinforcement or stiffening may be required. This may be combined with conventional stress analysis, if required, although the mechanical properties of GRP laminates are likely to be variable with lay-up technique employed, as well as the type of reinforcement used. Thus the specific figures for physical properties

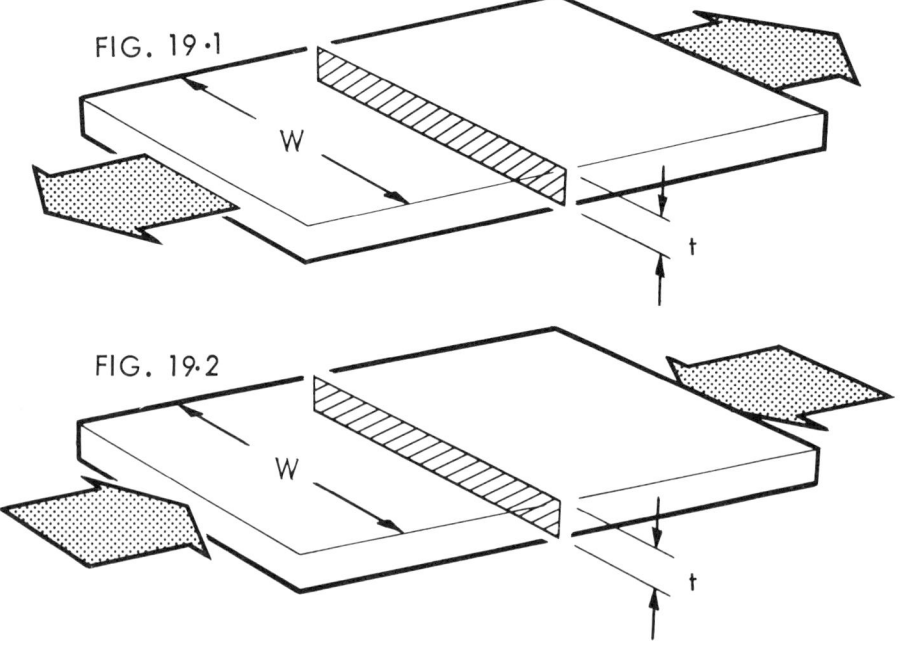

FIG. 19·1

FIG. 19·2

FIG. 19.3

FIG. 19.4    MOULDED-IN STIFFENERS

ADDED RIBS

given in the Appendix are not necessarily applicable directly to all GRP mouldings using the same reinforcement.

*Strength in Tension (Fig. 19.1)*

This is simply determined since the unit force or stress can be taken as uniformly distributed over the cross sectional area involved, i.e. width ($W$) times thickness ($t$). The resulting material stress is this

$$\text{Stress(p.s.i.)} = \frac{\text{Tensile Load (pounds)}}{W \times t}$$

where both $W$ and $t$ are in inches.

The tensile strength figures given in Table 1 can be used for assessing the performance of a typical laminate in pure tension, using a generous safety factor to allow for possible variations in glass:resin ratio and lay-up technique.

*Strength in Compression (Fig. 19.2)*

A similar formula applies where the material is stressed in pure compression, viz.

$$\frac{\text{Stress}}{\text{(p.s.i.)}} = \frac{\text{Compressive Load (pounds)}}{W \times t}$$

Lacking specific data on the compressive strength of GRP, the maximum

compressive strength can be taken as equal to the maximum tensile strength.

Almost without exception, GRP mouldings are relatively thin and thus when stressed in compression will tend to fail by bending long before the material is stressed to anywhere near its compressive limit—see Table 8. Under such conditions the material is put into shear.

The main call when designing GRP mouldings for withstanding compressive loads, therefore, is to provide adequate stiffness to resist buckling, and delay the point of ultimate failure which would otherwise result from the buckling developing high shear stresses.

Exactly similar requirements usually apply where the GRP moulding is subject to bending loads. Because of the relatively low elastic modulus, GRP mouldings tend to have large deflections under bending loads. Apart from the fact that this will also produce high shear forces, large deflections are usually undesirable. Again, therefore, the primary requirement is usually for extra stiffening or rigidity to be built into the design,

Flat panels can be stiffened by introducing flanged edges—*Fig. 19.3.* This has the effect of providing a supported rather than a free edge. Further stiffening can then be introduced by adding moulded in corrugations running in an appropriate direction, or directions, in the panel itself. Alternatively, separate stiffeners can be added to the panel after moulding (*Fig. 19.4*). Either method can provide entirely satisfactory results.

Some forms of separate stiffeners are shown in *Fig. 19.5*. Essentially these are "girder" sections formed in glass mat. A suitable core material is only

needed to enable such sections to be moulded onto the main moulding, with additional mat, and do not, or need not, contribute to the actual stiffening strength. Thus inexpensive lightweight materials may be used for stiffener cores or forms. This is a simpler method of adding separate stiffeners by moulding the GRP sections separately and then attaching them as finished mouldings.

An important point when adding stiffening ribs to a GRP moulding

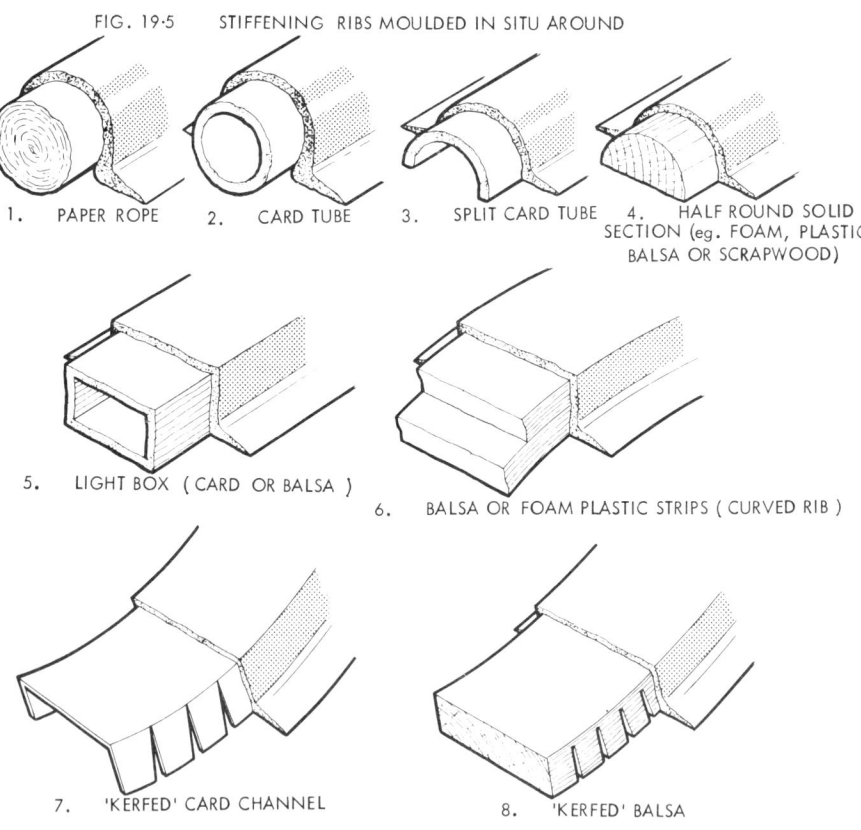

FIG. 19·5　STIFFENING RIBS MOULDED IN SITU AROUND

1.　PAPER ROPE　　2.　CARD TUBE　　3.　SPLIT CARD TUBE　　4.　HALF ROUND SOLID SECTION (eg. FOAM, PLASTIC BALSA OR SCRAPWOOD)

5.　LIGHT BOX ( CARD OR BALSA )

6.　BALSA OR FOAM PLASTIC STRIPS ( CURVED RIB )

7.　'KERFED' CARD CHANNEL

8.　'KERFED' BALSA

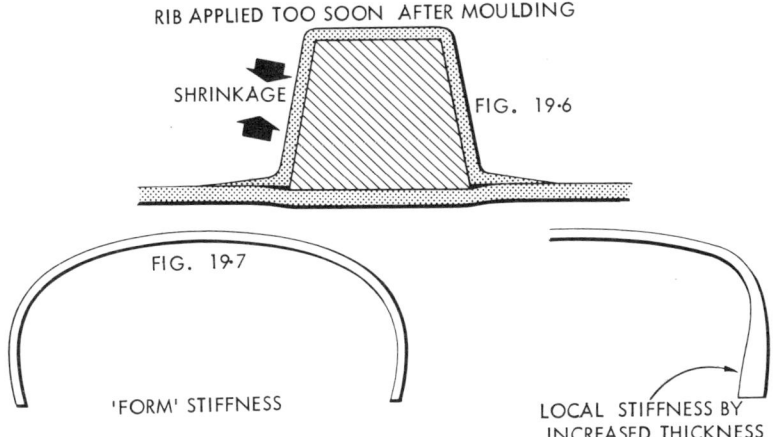

RIB APPLIED TOO SOON AFTER MOULDING

SHRINKAGE

FIG. 19·6

FIG. 19·7

'FORM' STIFFNESS

LOCAL STIFFNESS BY
INCREASED THICKNESS

is that the moulding should be set hard before the additional ribs are added. If glass on too soon, shrinkage can produce a distortion of the original moulding—see *Fig. 19.6*.

Stiffness can also be produced in other ways. The form of the moulding itself can provide inherent stiffness. Thus dished or curved cross-sections will be inherently stiffer than flat panels (*Fig. 19.7*). A further method of increasing stiffness locally is by building up additional layers of reinforcement in the regions where more stiffness is required. The two methods may be used together, and also in conjunction with moulded-in or added on stiffeners.

Where additional layers of reinforcement are used to increase local stiffness or strength, the increase in cross section produced should be progressive. Thus a concentrated build-up is bad, as it introduces a stress raiser or potential weakness at the

point of abrupt change of cross-section (*Fig. 19.8*). The additional thickness should be built up progressively, so that the change of section is gradual and reasonably uniform.

Complete GRP mouldings of large size are, of course, also commonly stiffened with internal ribs, frames or added members, often of wood, which are glassed in place in the finished moulding (*Fig. 19.9*). This is particularly applicable in the case of boat hull mouldings. Lugs, brackets and other sections may also be required, either glassed in, or bolted, screwed or riveted in place. The choice between bonding or mechanical fastening in such cases will depend largely on the nature of the fitting involved, its function, and the material from which it is made. Reinforcements to the moulding itself are invariably bonded in, to give maximum strength. Additional fittings are preferably bonded in, but may need further treatment.

FIG. 19·8 ADDITIONAL LAYERS OF REINFORCEMENT
GLASSED ON

POOR

GOOD

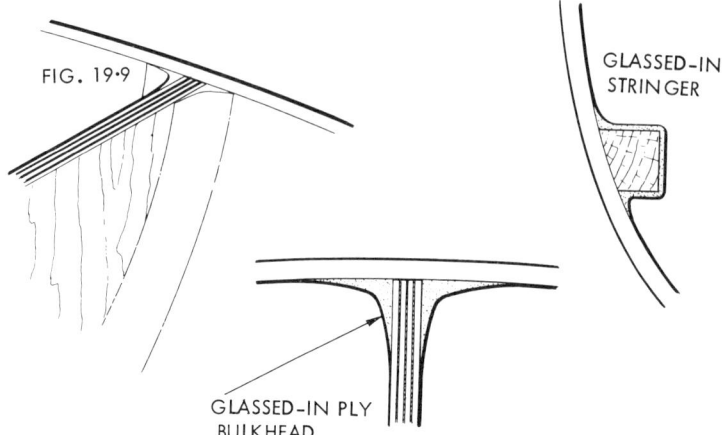

FIG. 19·9

GLASSED-IN STRINGER

GLASSED-IN PLY BULKHEAD

The successful use of polyester resin as an adhesive depends to a great extent on the nature of the surfaces to be joined, as well as the type of joint made. Polyester to polyester joints should have an ample overlap where practical, and to ensure a first-class bond, should be made before the resin has fully cured out in the parts to be joined. Up to six hours after the commencement of the hardening process at normal room temperatures is recommended for a complete adhesion. Too much reliance should not be placed on the resin for filling gaps, as it is naturally hard and brittle, and without glass reinforcing is liable to fracture.

Edge-to-edge joining of polyester sheets is not to be recommended, but if it is unavoidable, a good lap at the "back" of the laminate, formed of several layers of cloth or tape, can be used (*Fig. 19.10*). Joining other materials to polyester by means of the resin largely depends on the nature of the material. As the field of possible materials is so wide, the best approach is to try out a specimen piece and so discover whether the method would be satisfactory. Generally speaking any materials that have absorbent surfaces are no problem; hard polished surfaces such as glass, some plastics, and polished metal have, are not practical to stick with the resin, as under stress it will break away cleanly. The resin will adhere well to a roughened metal surface, as the surface gives a sufficient "key" for the resin to stick. Small metal brackets for joining other parts to a laminate are best moulded into the job as it is laid up. The brackets should have a sufficient flat area in contact with the resin, and should have several layers of cloth or mat, cut with a good overlap, laid over the bracket—see *Fig. 19.11*. Brackets moulded in, in this manner, will be amply strong and the thickening of the laminate locally will help to distribute the load.

  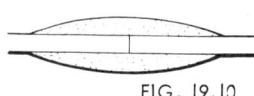

FIG. 19.10

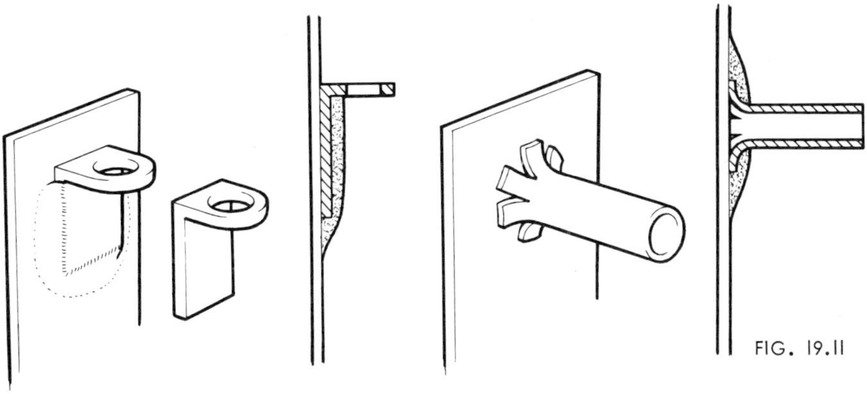

FIG. 19.11

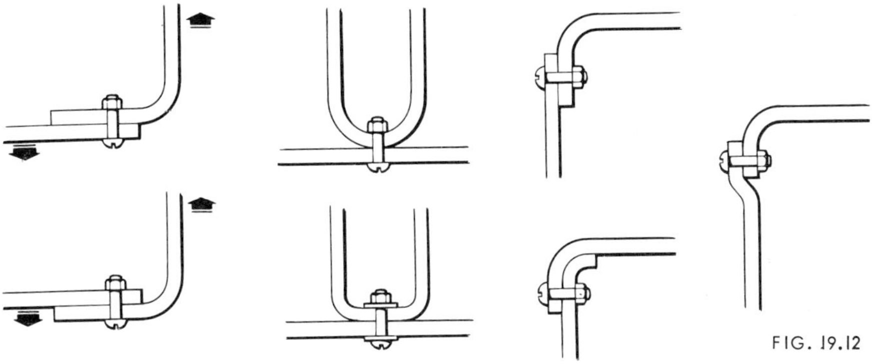

FIG. 19.12

Screws, rivets and bolts are used for joining laminates, with and without the addition of resin. The main consideration is to avoid putting too great a stress on the local area at the attachment joint. Washers under the heads help to spread the load, and reference to Table 7 giving minimum edge

*Table 7.* Edge Distances and Laminate Thickness

| Rivet Diameter | Laminate Thickness | Edge Distance (minimum) | Minimum Laminate Thickness for Countersinking |
|:---:|:---:|:---:|:---:|
| $\frac{3}{32}$ | $\frac{1}{8}$ | $\frac{5}{16}$ | ·080 |
| $\frac{1}{8}$ | $\frac{3}{16}$ | $\frac{3}{8}$ | ·090 |
| $\frac{5}{32}$ | $\frac{1}{4}$ | $\frac{1}{2}$ | ·100 |
| $\frac{3}{16}$ | over $\frac{1}{4}$ | $\frac{1}{2}$ | ·110 |

All measurements in inches

distances, hole sizes and thicknesses for countersinking, and to *Fig. 19.12*, will give some idea of the requirements in this direction.

When designing a moulding it is essential to think of the material as one of variable thickness and to determine the *amount of glass reinforcing* required in each part, or at specific points, bearing in mind the shape and rigidity required. In fact, if the moulding is rigid enough the other factors, such as tensile and compressive strength, resistance to shear, etc., are amply covered. It is easy to vary the thickness of the laminate at any point by adding extra layers of glass fibre. Obviously the best, lightest and most economic moulding can be produced in this way.

It is far better to specify the amount of glass fibre in any part of a moulding than to specify the *thickness* as, by varying the ratio of glass to resin, laminates of the same thickness can have vastly different strengths. By specifying the amount of glass, lam-inates of different thickness due to varying amounts of resin will then have similar strength figures.

## Sandwich Construction

Sandwich construction may be used where high stiffness is required, with minimum increase in weight. The 'sandwich' consists of two thin layers of glass resin on either side of a low density core material (*Fig. 19.13*). The stiffness is directly related to the total thickness of the sandwich and not only is this type of construction lighter than a thicker section of GRP, or a GRP skin with stiffeners added, but the total cost can be considerably reduced.

Suitable core materials include balsa, plastic foam, or honeycomb materials made of paper, glass fibre or metal. The latter group are generally excluded for ordinary work, because of high cost. Balsa is probably the best material where a high strength, completely durable sandwich is required.

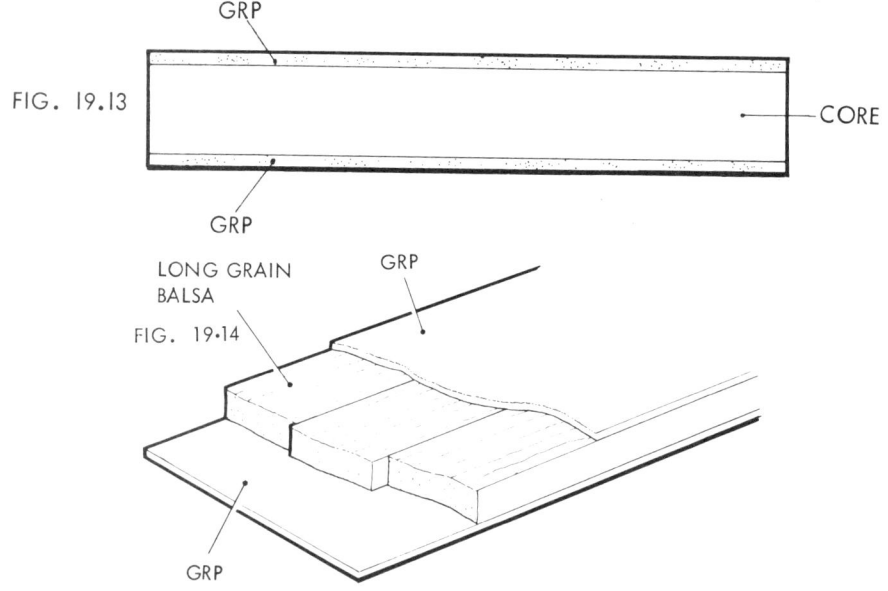

GRP

FIG. 19.13                                          CORE

GRP

LONG GRAIN BALSA

GRP

FIG. 19.14

GRP

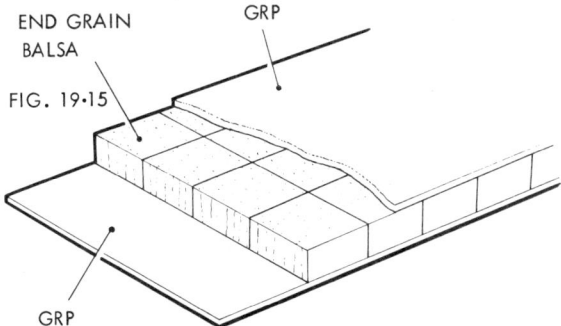

END GRAIN
BALSA

GRP

FIG. 19·15

GRP

It wets out well with the resin, which also serves to seal the surface of the wood and render it impervious to rot or waterlogging.

For producing flat sandwich panels, normal straight grain balsa can be employed, laid out in the form of sheets (*Fig. 19.14*). For curved panels end grain balsa is used, consisting of blocks about 2 in. square, cut to the required thickness (*Fig. 19.15*). These blocks can be laid individually to match almost any type of curve. End grain blocks can also be glued up to a thin scrim to form complete panels which can be draped in place, when laying up in the GRP moulding. In this case the scrim material may be terylene, nylon or glass fibre. The adhesive commonly used for gluing the individual blocks to the scrim is PVA, which is fully compatible with polyester resin.

The same technique can be applied to amateur constructions. The main difficulty here when using end grain balsa is a to ensure that the blocks are cut to exactly the same thickness, so that the core is of uniform thickness throughout the finished moulding. Individual blocks should be cut on a sawbench, or in a jig. Hand sawing, or attempting to face individual blocks down to the same thickness, is not practical.

The particular virtue of end grain balsa is its high compressive strength. The use of end grain balsa as a core material, therefore, not only increases the stiffness of the GRP moulding, but also its compressive strength and resistance to impact.

The added strength is directly related to the density of balsa used, which also affects the total weight of the laminate. Light density balsa should be selected for lightweight sandwich construction, as stiffness is largely independent of the core strength.

For the very lightest sandwich constructions, plastic foam is normally used for the core. All the foam materials have certain limitations as regards suitability and durability, however. Many tend to crumble or fail in use, with resulting loss of mechanical performance from the sandwich because of the break-up of the core. Only polyurethane foam and *unplasticised* rigid PVC foam can be used as core materials without pretreatment. Phenolic foam needs pretreatment with a thixotropic resin mixture before using to seal the surface and prevent penetration of lay-up resin into the foam. Whilst penetration is not necessarily harmful, it increases the weight of the sandwich, and also the cost, because of the greater quantity of resin absorbed.

Plasticised PVC foam needs sealing

before use as a core material as otherwise the resin may attack and soften the foam. Expanded polystyrene is rapidly attacked by the resin, but can be used if the surface is first fully sealed. Expanded rubber foam also needs to be sealed with shellac, or a special primer. Plastic foam materials which form a skin, either naturally, or after sealing, may also need their surface roughened before laying-up in order to ensure proper adhesion to the resin.

The stiffness of a sandwich panel can be calculated from the standard engineering formula

Deflection produced by load at mid-span $= P\dfrac{KL^3}{48}$

where $P =$ applied load
$L =$ distance between supports
$K = \dfrac{I}{bd^2}$

(see *Fig. 19.16*).

This formula ignores the contribution of the core, and is directly applicable in the case of plastic foam core materials where the core strength s negligible. Table 9 shows the relative stiffness of solid GRP and GRP sandwiches, calculated on this basis. The savings possible are obvious.

The true deflection, taking into account the shear modulus of the core

material is given by

$$\text{Deflection} = P\left[\frac{KL^3}{48} \times \frac{K_c L}{4}\right]$$

where $K_c = \dfrac{C}{bdGc}$

$Gc$ being the shear modulus of the core material.

This formula should be used to calculate deflections likely using a balsa core, which will give lower deflections for the same load than that given by the basic formula above. The shear modulus of balsa is given in Table 10.

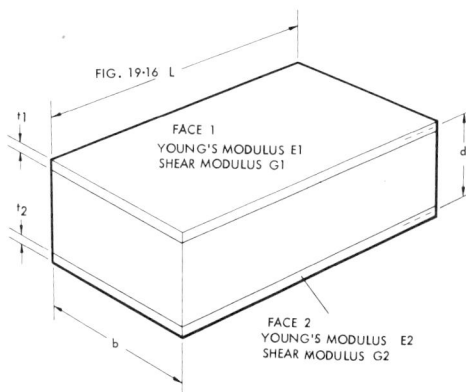

FIG. 19·16 L

FACE 1
YOUNG'S MODULUS E1
SHEAR MODULUS G1

FACE 2
YOUNG'S MODULUS E2
SHEAR MODULUS G2

---

*Design Nomograms*

The following two nomograms will be found very useful for estimating the type and thickness of glass fibre laminate required to compare in strength with either sheet steel or sheet aluminium. They were originated by the author and reproduced here by courtesy of *Design and Components in Engineering.*

**Nomogram 1. Comparable Strength with Sheet Steel**

Method of using the nomogram:

1. Find the connecting line from the strength quality being investigated to the type of laminate it is intended to use. Note the position of the terminal mark (semi-circle or rectangle) on the central line.

2. Lay a straight edge from the sheet

*Table 8*

Stiffness of typical GRP (chopped strand mat)

| Ounces of mat per sq. ft of laminate | Centre load to produce 0·1 in. deflection over 12 in. span | Load to break (pounds) |
|---|---|---|
| 2 | 3 | 30 |
| 3 | 10 | 60 |
| 4 | 22 | 90 |
| 5 | 44 | 130 |
| 6 | 66 | 170 |
| 7 | 87 | 220 |
| 8 | 108 | 275 |

*Table 9*

Comparative Stiffness of Solid and Sandwich GRP* (stiffness of 1 in. solid GRP = 100)

| Total thickness, inches | Solid GRP | Sandwich Construction | |
|---|---|---|---|
| | | $\frac{1}{16}$ in. GRP skins | $\frac{1}{8}$ in. GRP skins |
| $\frac{1}{16}$ | 0·018 | — | — |
| $\frac{1}{8}$ | 0·2 | — | — |
| $\frac{1}{4}$ | 1·6 | 1·3 | — |
| $\frac{3}{8}$ | 5·3 | 3·7 | 5 |
| $\frac{1}{2}$ | 12·5 | 7·3 | 10 |
| $\frac{3}{4}$ | 42 | 18 | 29 |
| 1 | 100 | 34 | 58 |
| $1\frac{1}{4}$ | 195 | 52 | 94 |
| $1\frac{1}{2}$ | 337 | 77 | 143 |
| $1\frac{3}{4}$ | 536 | 104 | 197 |
| 2 | 800 | 145 | 266 |

*Based on data by Fibreglass Ltd.

steel thickness scale through the terminal semi-circle (or extremes of the rectangle), and read off the appropriate laminate thickness range on the left-hand scale.

This nomogram gives the thickness of glass-fibre/polyester resin laminate required for comparable stiffness, flexural strength, shear strength, impact strength or tensile strength against steel sheet of known or specified thickness. Various forms of GF/polyester laminate construction are included, such as:

Unidirectional cloth (60% glass)—
CLOTH UD

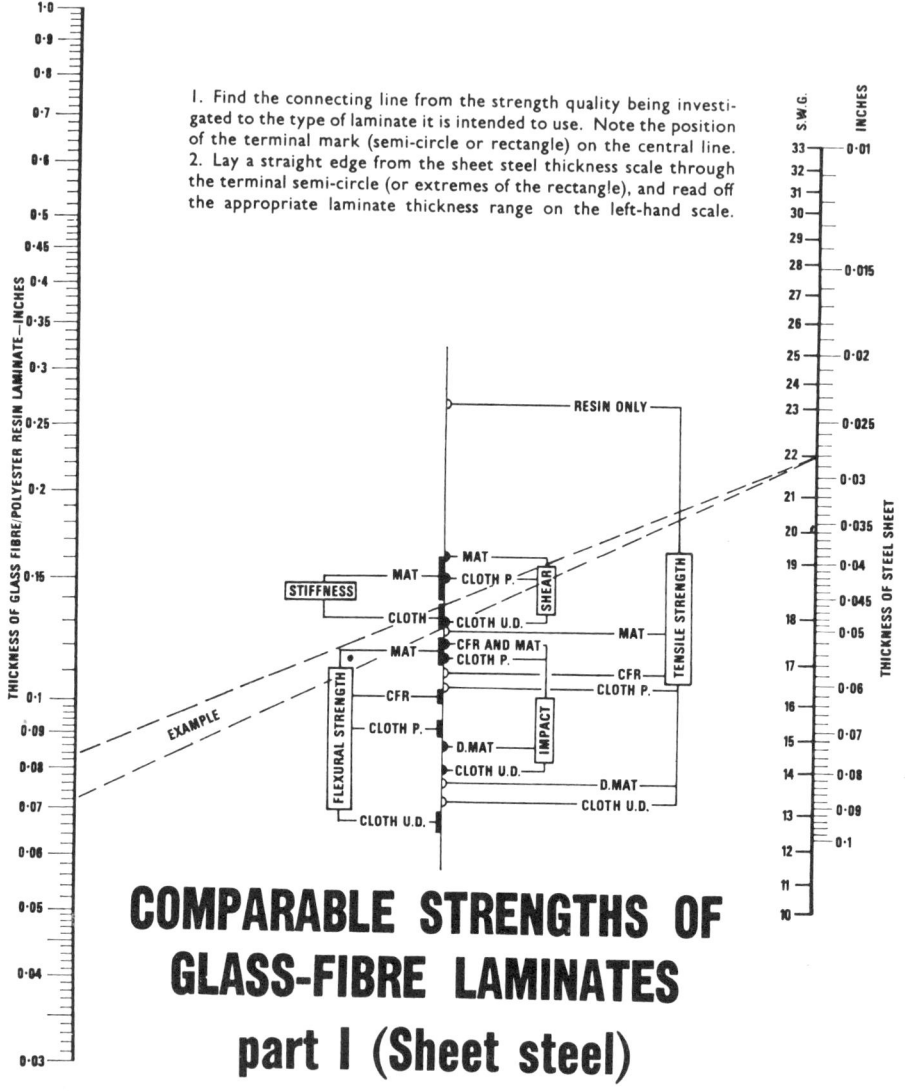

1. Find the connecting line from the strength quality being investigated to the type of laminate it is intended to use. Note the position of the terminal mark (semi-circle or rectangle) on the central line.
2. Lay a straight edge from the sheet steel thickness scale through the terminal semi-circle (or extremes of the rectangle), and read off the appropriate laminate thickness range on the left-hand scale.

# COMPARABLE STRENGTHS OF GLASS-FIBRE LAMINATES part I (Sheet steel)

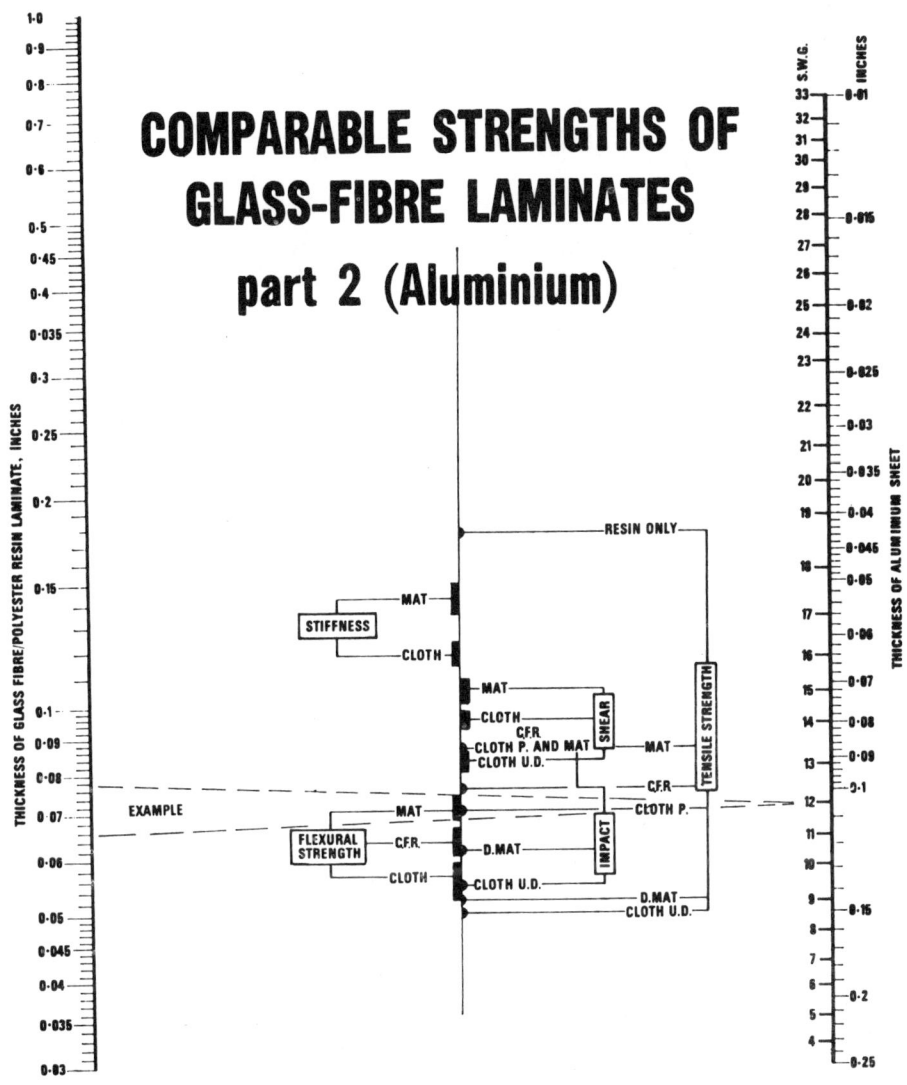

Typical plain weave cloth (60 % glass)—CLOTH
Typical chopped strand mat (35–45 % glass)—MAT
Chopped fibre rovings (55 % glass)—CFR
Diamond mat (unidirectional) (50–55 % glass)—D.MAT

To use the nomogram, connect the appropriate steel sheet thickness with a straight line through the appropriate point or two lines through the extremes of the appropriate mark on the centre line and continue to cut the left-hand scale. Read off equivalent GF/polyester laminate thickness in inches. Note that the constructions designated on the centre lines are specific to the property to which they are connected, for example "Stiffness", "Flexural Strength", etc.

*Example:* to find the thickness of GF/polyester laminate required to give similar stiffness to 22 swg steel sheet. **Ans: 0·07 in. to 0·084 in.**

Thickness based on the nomogram solutions are for typical good class laminates with a glass content of the order expressed above. Solutions are approximate rather than exact, since a considerable number of variables are involved in practical construction which could affect the mechanical performance of the final laminate

The nomogram can also be used the other way round to determine the approximate equivalent thickness in steel of a GF/polyester laminate of given thickness and particular construction.

**Nomogram 2. Comparable strength with Sheet Aluminium**

This nomogram relates equivalent thicknesses for approximately equal stiffness, flexural strength, shear strength, impact strength and tensile strength of GF/polyester laminates of various construction and typical alu-

minium sheet. Mechanical properties of the aluminium sheet are equivalent to half hard temper. Various forms of GF/polyester laminates are included, such as:

Unidirectional cloth (60 % glass)—CLOTH UD
Typical Plain Weave cloth (60 % glass)—CLOTH
Typical chopped strand mat (35–45 % glass)—MAT
Chopped fibre rovings (55 % glass)—CFR
Unidirectional Diamond Mat (50–55 % glass)—D. MAT

To use the nomogram, connect the appropriate aluminium sheet thickness with a straight line through the appropriate mark (or two lines through the extreme ends of the appropriate mark) on the centre line and project to cut the left hand scale. Read off equivalent GF/polyester laminate thickness. Note that the constructions designated on the centre line are specific to the property to which they are connected—e.g. "Stiffness",' 'Flexural Strength", etc.

*Example:* to find the thickness of GF/polyester resin laminate in chopped strand mat to give a similar flexural strength to 12 swg half-hard aluminium sheet. **Ans: 0·066 in to 00·78 in.**

Thicknesses given by this nomogram are based on typical good class laminates with a glass content of the order given above. It will be appreciated that solutions are approximate rather than exact since a considerable number of variables are involved in practical construction which could affect the mechanical properties of the final laminate.

The nomogram can also be worked the other way round to find the approximate equivalent thickness in half hard aluminium sheet of a GF/polyester laminate of given thickness and particular construction.

# APPENDICES

MECHANICAL strength figures are directly related to the amount of glass in any laminate, and also to the lay-up technique, etc. They cannot be related to "typical" or "average" performance as with other standard materials like metals and woods. For critical mouldings, maximum strength figures are obtained by tight specification of materials and fabrication technique, but for all normal applications ultimate strength is seldom an important factor.

The following data, which give a useful guide to strength figures obtainable, relate to specific test specimens of average performances which could be expected, utilising sound production techniques.

## Typical Physical Properties of Glass Plastic Laminates with Polyester Resin

| Property | Cloth | Chopped Strand Mat | Rovings |
|---|---|---|---|
| Specific Gravity | 1·65–1·7 | 1·55–1·6 | 1·8–1·9 |
| Glass Content, % | 55 | 30 | 70 |
| Tensile Strength, p.s.i. | 40–50,000 | 15–25,000 | 80–120,000 |
| Compressive Strength, p.s.i. | 30–35,000 | 15–25,000 | 60–70,000 |
| Bending Strength, p.s.i. | 50–60,000 | 25–30,000 | 120–150,000 |
| Impact Strength, ft. lb./in² | 20–25 | 18–20 | 60–70 |
| Coefficient of Linear Expansion ($\times 10^{-6}$ per⁹C) | 10 | 12 | 15–17 |

*Table 10*

Properties of Balsa in Bending

| Density lb./cu. ft | Static bending | | |
|---|---|---|---|
| | Modulus of elasticity p.s.i. $\times 10^4$ | Modulus of rupture p.s.i. | Elastic limit* |
| 6 | 26·7 | 1,420 | |
| 8 | 43·2 | 2,480 | |
| 9 | 52·0 | 3,020 | Varies from |
| 10 | 60·5 | 3,540 | 75–95% |
| 11 | 68·9 | 4,080 | Modulus of |
| 12 | 77·5 | 4,620 | Rupture |
| 14 | 94·0 | 5,700 | |
| 16 | 111·0 | 6,760 | |

*Based on data by Fibreglass Ltd.

## Approximate Weights and Thickness of Laminates

| Material | No. of layers | Glass weight oz/sq.ft | Resin weight oz/sq.ft | Glass: Resin (%) | Laminate weight oz/sq.ft | Laminate weight lb/sq.yd | Approx. fibreglass in. |
|---|---|---|---|---|---|---|---|
| 1 oz Glass Mat | 1 | 1 | 2·5 | 30 | 3·5 | 2·0 | $\frac{1}{32}$ |
| | 2 | 2 | 4 | 30 | 6 | 3·9 | $\frac{1}{16}$ |
| | 3 | 3 | 6 | 30 | 9 | 5·1 | $\frac{3}{32}$ |
| 1½ oz Glass Mat | 1 | 1½ | 3·5 | 30 | 5 | 2·8 | $\frac{3}{64}$ |
| | 2 | 3 | 6 | 30 | 9 | 5·1 | $\frac{3}{32}$ |
| | 3 | 4½ | 9 | 30 | 13·5 | 7·6 | $\frac{9}{64}$ |
| 2 oz Glass Mat | 1 | 2 | 4 | 30 | 6 | 3·4 | $\frac{1}{16}$ |
| | 2 | 4 | 8 | 30 | 12 | 6·8 | $\frac{1}{8}$ |
| | 3 | 6 | 12 | 30 | 18 | 10·2 | $\frac{3}{16}$ |
| 0·015 in. scrim | 1 | 1 | 2 | 30 | 3 | 1·7 | 0·035 |
| | 2 | 2 | 3 | 30 | 5 | 2·8 | 0·050 |
| | 4 | 4 | 5·5 | 40 | 9·5 | 5·4 | 0·085 |
| Cloth (0·009 in. close weave) | 1 | 0·8 | 0·6 | 60 | 1·4 | 0·8 | 0·015 |
| | 2 | 1·6 | 1·2 | 60 | 2·8 | 1·6 | 0·030 |
| | 4 | 3·2 | 2·4 | 60 | 5·6 | 3·2 | 0·060 |
| Cloth (0·010 in. open weave) | 1 | 0·8 | 1·2 | 40 | 2·0 | 1·1 | 0·020 |
| | 2 | 1·6 | 2·4 | 40 | 4·0 | 2·3 | 0·035 |
| | 4 | 3·2 | 3·6 | 45 | 6·8 | 3·8 | 0·055 |
| Surface tissue | 1 | 0·075 | 1·0 | — | — | — | — |

# Comparative Material Performance

| Material | Glass Content $\%$ by weight | Glass Content $\%$ by volume | Specific Gravities | Density lb./cu. in. | Thermal Coefficient of Expansion in/in/°F $\times 10^{-6}$ | Tensile Strength* lbf/in² $\times 10^{-3}$ | Tensile Modulus lbf/in² $\times 10^{-6}$ | Compressive Strength lbf/in² $\times 10^{-3}$ | Flexural Strength lbf/in² $\times 10^{-3}$ |
|---|---|---|---|---|---|---|---|---|---|
| **FIBREGLASS REINFORCED:** | | | | | | | | | |
| Uni directional Roving wound epoxide.. | 60–90 | 40–80 | 1·7–2·2 | ·061–·079 | 2–6 | 80–250 | 4–9 | 45–70 | 100–270 |
| extrusion polyester | 50–75 | 32–59 | 1·6–2·0 | ·057–·072 | 3–8 | 60–170 | 3–6 | 30–70 | 100–180 |
| Bi directional Fabric satin weave polyester .. | 50–70 | 32–52 | 1·6–1·9 | ·057–·068 | 5–6 | 35–58 | 2–3·5 | 30–40 | 50–75 |
| woven roving polyester | 45–60 | 28–41 | 1·5–1·8 | ·054–·065 | 6–9 | 32–50 | 1·8–2·4 | 14–20 | 30–43 |
| Random Mat/Spray Up preform polyester | 25–50 | 14–32 | 1·4–1·6 | ·052–·058 | 10–18 | 10–24 | 0·8–1·8 | 18–30 | 20–45 |
| hand lay up polyester } spray up polyester } | 25–40 | 14–24 | 1·4–1·5 | ·052–·054 | 12–20 | 9–20 | 0·8–1·6 | 18–25 | 20–40 |
| Moulding Compound Dough moulding compound polyester | 6–26 | 10–24 | 1·8–2·0 | ·065–·073 | 13–19 | 5–10 | 1·6–2·0 | 20–26 | 6–26 |
| glass/nylon | 20–40 | | 1·3–1·5 | ·048–·055 | 7–18 | 17–28 | 0·8–2·0 | 15–24 | 21–40 |
| **THERMOPLASTICS** | | | | | | | | | |
| Nylon | | | 1·14 | 0·042 | 55–63 | 11·5 | 0·2–0·4 | 5–13 | 8–15 |
| Polyethylene (high density) | | | 0·96 | 0·035 | 61–72 | 4·4 | 0·08–0·15 | 2·4 | 2–3 |
| Polypropylene .. | | | 0·90 | 0·033 | 55–111 | 5·7 | 0·15–0·25 | 8·5–10 | 4–5 |
| Polystyrene (high impact) | | | 1·08 | 0·040 | 22–56 | 6·5 | 0·5 | 16 | 3–5 |
| **METALS** | | | | | | | | | |
| Aluminium | | | 2·7 | 0·10 | 12–13 | 11–26 | 10 | 12 | 32 |
| Mild steel | | | 7·8 | 0·28 | 6–8 | 34 | 30 | 28 | 28 |
| Stainless steel .. | | | 7·92 | 0·29 | 9–10 | 29–45 | 28 | 30 | 28 |

*Yield stress for metals, ultimate for other materials.

*(Courtesy of Fibreglass Ltd.)*

# The New Glass Fibre Book

## Temperature Conversion Tables

| C. | | F. | C. | | F. | C. | | F. |
|---|---|---|---|---|---|---|---|---|
| −1·11 | 30 | 86·0 | 12·2 | 54 | 129·2 | 26·1 | 79 | 174·2 |
| −0·56 | 31 | 87·8 | 12·8 | 55 | 131·0 | 26·7 | 80 | 176·0 |
| 0 | 32 | 89·6 | 13·3 | 56 | 132·8 | 27·2 | 81 | 177·8 |
| 0·56 | 33 | 91·4 | 13·9 | 57 | 134·6 | 27·8 | 82 | 179·6 |
| 1·11 | 34 | 93·2 | 14·4 | 58 | 136·4 | 28·3 | 83 | 181·4 |
| 1·67 | 35 | 95·0 | 15·0 | 59 | 138·2 | 28·9 | 84 | 183·2 |
| 2·22 | 36 | 96·8 | 15·6 | 60 | 140·0 | 29·4 | 85 | 185·0 |
| 2·78 | 37 | 98·6 | 16·1 | 61 | 141·8 | 30·0 | 86 | 186·8 |
| 3·33 | 38 | 100·4 | 16·7 | 62 | 143·6 | 30·6 | 87 | 188·6 |
| 3·89 | 39 | 102·2 | 17·2 | 63 | 145·4 | 31·1 | 88 | 190·4 |
| 4·44 | 40 | 104·0 | 17·8 | 64 | 147·2 | 31·7 | 89 | 192·2 |
| 5·00 | 41 | 105·8 | 18·3 | 65 | 149·0 | 32·2 | 90 | 194·0 |
| 5·56 | 42 | 107·6 | 18·9 | 66 | 150·8 | 32·8 | 91 | 195·8 |
| 6·11 | 43 | 109·4 | 19·4 | 67 | 152·6 | 33·3 | 92 | 197·6 |
| 6·67 | 44 | 111·2 | 20·0 | 68 | 154·4 | 33·9 | 93 | 199·4 |
| 7·22 | 45 | 113·0 | 20·6 | 69 | 156·2 | 34·4 | 94 | 201·2 |
| 7·78 | 46 | 114·8 | 21·1 | 70 | 158·0 | 35·0 | 95 | 203·0 |
| 8·33 | 47 | 116·6 | 21·7 | 71 | 159·8 | 35·6 | 96 | 204·8 |
| 8·89 | 48 | 118·4 | 22·2 | 72 | 161·6 | 36·1 | 97 | 206·6 |
| 9·44 | 49 | 120·2 | 22·8 | 73 | 163·4 | 36·7 | 98 | 208·4 |
| 10·0 | 50 | 122·0 | 23·3 | 74 | 165·2 | 37·2 | 99 | 210·2 |
| 10·6 | 51 | 123·8 | 23·9 | 75 | 167·0 | 37·8 | 100 | 212·0 |
| 11·1 | 52 | 125·6 | 24·4 | 76 | 168·8 | | | |
| 11·7 | 53 | 127·4 | 25·0 | 77 | 170·6 | | | |
| | | | 25·6 | 78 | 172·4 | | | |

To use this table, enter the known value of temperature on the centre (bold) scale and read off the appropriate equivalent under the respective Fahrenheit or Centigrade column. For example, to convert 68 degrees F. into degreec C., read off equivalent against 68 on centre scale under the Centigrade column. Answer 20·0 degrees C.

## Material Suppliers

The following are the main suppliers of GRP materials, kits, packs, etc., which are available from do-it-yourself shops, ships' chandlers, motor accessory shops, etc.

*Automobile Plastics Ltd.*, 7 Henry Road, New Barnet, Herts.
*Bondaglass Ltd.*, West Wickham, Kent.

*Griffin and George Ltd.*, Ealing Road, Alperton, Wembley, Middx.

*Holts* (Douglas Holt, Ltd.), New Addington, Surrey.

*Isopon Inter Chemicals*, Derbyshire House, St. Chad's Street, London W.C. 1.

*Silver Dee Plastics Ltd.*, Staveley, Chesterfield, Derby.

*Strand Glass Co. Ltd.*, 109 High Street, Brentford, Middlesex. 980 Stockport Road, Manchester, 19. 72 London Road, Southampton.

*Alec Tiranti Ltd.*, 72 Charlotte St., London, W.1. P2AJ.

*Trylon Ltd.*, Thrift Street, Wollaston, Northants.

The following are suppliers of basic materials to the trade. They do not normally supply in small quantities, but may be contacted for information as to where specialised products may be obtained for amateur work or small scale productions.

*Polyester Resins*

Astor Boisselier & Lawrence, Ltd., West Drayton, Middlesex.

Automobile Plastics Ltd., 7 Henry Road, New Barnet, Herts.

Bakelite Xynolite Ltd., PO Box 103, Abbey Lane, Leicester.

Beck, Koller and Co. (England) Ltd., Speke, Liverpool 14.

B.I.P. Chemicals Ltd., PO Box 6, Warley, Worcs.

Scott Bader & Co. Ltd., Wellingborough, Northants.

J. M. Steel & Co. Ltd., 18–24 Paradise Row, Richmond, Surrey.

*Fillers*

Witco Chemical Co. Ltd., Bush House, Aldwych, London, W.C.2.

H. & A. Watson & Co. Ltd., 448, Derby House, Liverpool 2.

John & E. Sturge Ltd., Wheeley's Lane, Birmingham, 15

Croxton & Garry Ltd., 16/18 High Street, Kingston-upon-Thames, Surrey.

*Plaster of Paris*

Cafferata & Co. Ltd., Newark, Notts.

*Pigments and Pigment Pastes*

Ferro Enamels Ltd., Wombourn, Wolverhampton.

West & Senior Ltd., Chatsworth Mills, William Street, Gorton, Manchester 18.

Harrisons & Sons (Hanley) Ltd., Garth Street, Hanley, Stoke-on-Trent.

Llewellyn Ryland Ltd., Balsall Heath Works, Hadem Street, Birmingham 12.

Docker Brothers, P.O. Box 359, Rotten Park Street, Birmingham 16.

Blundell, Spence & Co. Ltd., Industrial Division, Trading Estate, Slough, Bucks.

*Fire-retarding Agents*

Chlorinated Hydrocarbons

Imperial Chemical Industries Ltd., Nobel House, Buckingham Gate, London, S.W.1.

CERECLOR

Monsanto Chemicals Ltd., Monsanto House, 10/18, Victoria Street, London, S.W.1.

AROCLOR

Antimony Oxide

Associated Lead Manufacturers Ltd., 14, Finsbury Circus, London, E.C.2.

*Open Rollers*

Shawcraft Limited, 69, Rockingham Road, Uxbridge, Middlesex.

Paytham Engineering Co., Lombard Street, Birmingham.

*Glass Reinforcing Materials*

Fibreglass Limited, Ravenhead, St. Helens, Lancs.

Fothergill & Harvey Ltd., Harvester House, 37, Peter Street, Manchester 2.

Deeside Fibres Limited, Prince Consort House, 27–22, Albert Embankment, London, S.E.1.

Marglass Limited, Sherborne, Dorset.

Turner Brothers Asbestos Co. Ltd., 14, Finsbury Circus, London, E.C.2.

*'Cellophane'*—non-moisture proof 600

British Cellophane Ltd., 12/13, Conduit Street, London, W. 1.

*Wax Release Agents*

Simoniz (England) Ltd., 125, High Holborn, London, W.C.1.

S. E. Johnson Ltd., "Car Plate" High Street, West Drayton, Middlesex.

*Barrier Creams*

Rozalex Ltd., 10, Norfolk Street, Manchester.

Scientific Pharmacals Ltd., 1, Eden Street, Hampstead Road, London, N.W.1.